6th International Workshop on New Computational Methods for Inverse Problems (NCMIP 2016)

AA006333

Journal of Physics: Conference Series Volume 756

Cachan, France
20 May 2016

ISBN: 978-1-5108-3106-3
ISSN: 1742-6588

Printed from e-media with permission by:

Curran Associates, Inc.
57 Morehouse Lane
Red Hook, NY 12571

Some format issues inherent in the e-media version may also appear in this print version.

Printed by Curran Associates, Inc. (2016)

For permission requests, please contact the Institute of Physics
at the address below.

Institute of Physics
Dirac House, Temple Back
Bristol BS1 6BE UK

Phone: 44 1 17 929 7481
Fax: 44 1 17 920 0979

techtracking@iop.org

Additional copies of this publication are available from:

Curran Associates, Inc.
57 Morehouse Lane
Red Hook, NY 12571 USA
Phone: 845-758-0400
Fax: 845-758-2633
Email: curran@proceedings.com
Web: www.proceedings.com

Table of contents

Volume 756

6th International Workshop on New Computational Methods for Inverse Problems
20 May 2016, Cachan, France

Accepted papers received: 20 September 2016
Published online: 7 October 2016

Preface

Papers

Foreword

This volume of Journal of Physics: Conference Series is dedicated to the scientific contributions presented during the 6th International Workshop on New Computational Methods for Inverse Problems, NCMIP 2016 (http://complement.farman.ens-cachan.fr/NCMIP_2016.html). This workshop took place at Ecole Normale Supérieure de Cachan, on May 20, 2016. The prior editions of NCMIP also took place in Cachan, France, firstly within the scope of ValueTools Conference, in May 2011, and secondly at the initiative of Institut Farman, in May 2012, May 2013, May 2014 and May 2015.

The New Computational Methods for Inverse Problems (NCMIP) workshop focused on recent advances in the resolution of inverse problems. Indeed, inverse problems appear in numerous scientific areas such as geophysics, biological and medical imaging, material and structure characterization, electrical, mechanical and civil engineering, and finances. The resolution of inverse problems consists in estimating the parameters of the observed system or structure from data collected by an instrumental sensing or imaging device. Its success firstly requires the collection of relevant observation data. It also requires accurate models describing the physical interactions between the instrumental device and the observed system, as well as the intrinsic properties of the solution itself. Finally, it requires the design of robust, accurate and efficient inversion algorithms. Advanced sensor arrays and imaging devices provide high rate and high volume data; in this context, the efficient resolution of the inverse problem requires the joint development of new models and inversion methods, taking computational and implementation aspects into account. During this one-day workshop, researchers had the opportunity to bring to light and share new techniques and results in the field of inverse problems.

The topics of the workshop were: algorithms and computational aspects of inversion, Bayesian estimation, Kernel methods, learning methods, convex optimization, free discontinuity problems, metamodels, proper orthogonal decomposition, reduced models for the inversion, non-linear inverse scattering, image reconstruction and restoration, and applications (bio-medical imaging, non-destructive evaluation...).

NCMIP 2016 was a one-day workshop held in May 2016 which attracted around seventy attendees. Each of the submitted papers has been reviewed by two reviewers. There have been eleven accepted papers. In addition, three international speakers were invited to present a longer talk.

The workshop was supported by Institut Farman (ENS Cachan, CNRS) and endorsed by the following French research networks: GDR ISIS, GDR MIA, GDR MOA, GDR Ondes. The program committee acknowledges the following research laboratories: CMLA, LMT, LURPA and SATIE.

Eric Vourc'h and Thomas Rodet

Journal of Physics: Conference Series **756** (2016) 011001 doi:10.1088/1742-6596/756/1/011001

Workshop co-chairs:

Eric Vourc'h, SATIE laboratory, Ecole Normale Supérieure de Cachan, CNRS, France
Thomas Rodet, SATIE laboratory, Ecole Normale Supérieure de Cachan, CNRS, France

Technical program committee:

Alexandre Baussard, ENSTA Bretagne, Lab-STICC, France
Marc Bonnet, ENSTA, ParisTech, France
Laure Blanc-Féraud, I3S laboratory and INRIA Nice Sophia-Antipolis, France
Antonin Chambolle, CMAP, Ecole Polytechnique, CNRS, France
Oliver Dorn, School of Mathematics, University of Manchester, UK
Cécile Durieu, SATIE, ENS Cachan, CNRS, France
Laurent Fribourg, LSV, ENS Cachan, CNRS, France
Jerôme Idier, IRCCyN Laboratory, Ecole Centrale de Nantes, France
Pierre-Yves Joubert, IEF, Paris-Sud University, CNRS, France
Marc Lambert, Geeps Laboratory, CNRS, CentraleSupElec, Paris-Sud University, France
Giacomo Oliveri, eledia research center/eledia@L2S group, University of Trento, Italy
Dominique Lesselier, L2S Laboratory, CNRS, CentraleSupElec, Paris-Sud University, France
Matteo Pastorino, DIBE, University of Genoa, Italy
Gabriel Peyré, Ceremade laboratory, University of Paris Dauphine, France
Anthony Quinn, Trinity College Dublin, Ireland
Christian Rey, Safran Tech, France
Marco Salucci, eledia research center/eledia@L2S group, University of Trento, Italy
Jean-Yves Tourneret, IRIT laboratory, ENSEEIHT (Toulouse), France

Local chairs:

Virginie Pauchont, CMLA Laboratory, ENS Cachan, Paris-Sud University, CNRS, France
Sophie Abriet, SATIE Laboratory, ENS Cachan, Paris-Sud University, CNRS, France
Béatrice Bacquet, SATIE Laboratory, ENS Cachan, Paris-Sud University, CNRS, France
Aurore Gracia, SATIE Laboratory, ENS Cachan, Paris-Sud University, CNRS, France

Invited speakers

Laure Blanc-Féraud, I3S Laboratory, CNRS UMR 6070, Sophia Antipolis, France.
Michel Defrise, Department of Nuclear Medicine, Vrije Universiteit Brussel, Belgium.
Houssem Haddar, INRIA, CMAP, Ecole Polytechnique, Palaiseau, France.

Journal of Physics: Conference Series **756** (2016) 011002 doi:10.1088/1742-6596/756/1/011002

Peer review statement

All papers published in this volume of *Journal of Physics: Conference Series* have been peer reviewed through processes administered by the proceedings Editors. Reviews were conducted by expert referees to the professional and scientific standards expected of a proceedings journal published by IOP Publishing.

On the constrained minimization of smooth Kurdyka–Łojasiewicz functions with the scaled gradient projection method

Marco Prato[1], **Silvia Bonettini**[2], **Ignace Loris**[3], **Federica Porta**[2] and **Simone Rebegoldi**[1]

[1] Dipartimento di Scienze Fisiche, Informatiche e Matematiche, Università di Modena e Reggio Emilia, Via Campi 213/b, 41125 Modena, Italy
[2] Dipartimento di Matematica e Informatica, Università di Ferrara, Via Saragat 1, 44122 Ferrara, Italy
[3] Département de Mathématique, Université Libre de Bruxelles, Boulevard du Triomphe, 1050 Bruxelles, Belgium

E-mail: marco.prato@unimore.it

Abstract. The scaled gradient projection (SGP) method is a first-order optimization method applicable to the constrained minimization of smooth functions and exploiting a scaling matrix multiplying the gradient and a variable steplength parameter to improve the convergence of the scheme. For a general nonconvex function, the limit points of the sequence generated by SGP have been proved to be stationary, while in the convex case and with some restrictions on the choice of the scaling matrix the sequence itself converges to a constrained minimum point. In this paper we extend these convergence results by showing that the SGP sequence converges to a limit point provided that the objective function satisfies the Kurdyka–Łojasiewicz property at each point of its domain and its gradient is Lipschitz continuous.

1. The scaled gradient projection method

The scaled gradient projection algorithm (SGP) [4] belongs to the class of first-order methods designed to solve any constrained optimization problem of the form

$$\min_{x\in\mathbb{R}^n} \Psi(x) \equiv f(x) + \iota_\Omega(x), \tag{1}$$

where f is a continuously differentiable function and ι_Ω is the indicator function of the nonempty, closed and convex set Ω. In particular, the $(k+1)$–th SGP iteration is computed as

$$x^{(k+1)} = x^{(k)} + \lambda_k d^{(k)} = x^{(k)} + \lambda_k(\underbrace{P_{\Omega,D_k^{-1}}(x^{(k)} - \alpha_k D_k \nabla f(x^{(k)}))}_{=:y^{(k)}} - x^{(k)}),$$

where α_k is a scalar steplength parameter, D_k is a symmetric positive definite matrix and $P_{\Omega,D_k^{-1}}(\cdot)$ is the projection onto Ω associated to the norm induced by D_k^{-1} [4]. The step along the descent direction $d^{(k)}$ is performed by means of the linesearch parameter $\lambda_k = \delta^{m_k}$, where $\delta \in (0,1)$ and m_k is the smallest non-negative integer such that the monotone Armijo condition

$$f(x^{(k)} + \lambda_k d^{(k)}) \leq f(x^{(k)}) + \beta\lambda_k \nabla f(x^{(k)})^T d^{(k)} \tag{2}$$

is satisfied for a fixed value of the parameter $\beta \in (0,1)$.
The SGP method is a variable metric forward-backward algorithm [11, 12] which has been exploited in the last years for the solution of different real-world inverse problems [5, 6, 7, 16, 17, 18]. The main difference between SGP and the standard forward-backward schemes is the presence of two independent parameters α_k and λ_k with a complete different role: while the last one is automatically computed with the Armijo condition (2) to guarantee the sufficient decrease of the objective function, the first one can be chosen to improve the actual convergence rate of the method, exploiting thirty years of literature in numerical optimization [3, 13, 19, 20]. We also remark that, unlike the schemes presented e.g. in [2, Sect. 5], in SGP the explicit knowledge of the gradient Lipschitz constant is not required.
The general convergence result on the SGP sequence states that all its limit points are stationary for problem (1), provided that both the steplength α_k and the eigenvalues of D_k are chosen in prefixed positive intervals $[\alpha_{\min}, \alpha_{\max}]$ and $[\frac{1}{\mu}, \mu]$, respectively [4]. Convergence of the sequence to a minimum point of (1) has recently been proved for convex objective functions by choosing suitable adaptive bounds for the eigenvalues of the scaling matrices [8]. In the following, we will consider a modified version of SGP in which, at each iteration $k \in \mathbb{N}$, we compute

$$x^{(k+1)} = \begin{cases} y^{(k)} & \text{if } \ f(y^{(k)}) < f(x^{(k)} + \lambda_k d^{(k)}) \\ x^{(k)} + \lambda_k d^{(k)} & \text{otherwise.} \end{cases} \tag{3}$$

so that the sequence $\{f(x^{(k)})\}_{k\in\mathbb{N}}$ is forced to assume lower values than it originally would with SGP. By doing so, the new sequence satisfies the condition $f(x^{(k+1)}) \leq f(x^{(k)} + \lambda_k d^{(k)})$, which still guarantees the stationarity of the limit points [8]. The aim of this paper is to prove the convergence of this modified SGP scheme if the (nonconvex) objective function Ψ satisfies the Kurdyka–Łojasiewicz (KL) property [15, 14], which holds true for most of the functions commonly used in inverse problems as p norms, Kullback-Leibler divergence and indicator functions of box plus equality constraints. A generalization of the proposed method and the related convergence proof to the minimization of the sum of smooth (nonconvex) and convex (nonsmooth) KL functions is in progress [10].

2. Preliminary results

Let $\Psi : \mathbb{R}^n \longrightarrow \mathbb{R}\cup\{+\infty\}$ be a proper, lower semicontinuous function. For $-\infty < \eta_1 < \eta_2 \leq +\infty$, let us set $[\eta_1 < \Psi < \eta_2] = \{z \in \mathbb{R}^n : \eta_1 < \Psi(z) < \eta_2\}$. Moreover, we denote with $\partial\Psi(z)$ the subdifferential of Ψ at $z \in \mathbb{R}^n$ and with $\mathrm{dist}(z, \Omega)$ the distance between a point z and a set $\Omega \subset \mathbb{R}^n$. The function Ψ is said to have the KL property at $\overline{z} \in \mathrm{dom}\partial\Psi := \{z \in \mathbb{R}^n : \partial\Psi(z) \neq \emptyset\}$ if there exist $\eta \in (0, +\infty]$, a neighborhood U of $\overline{z}$ and a continuous concave function $\varphi : [0, \eta) \longrightarrow [0, +\infty)$ such that $\varphi(0) = 0$, φ is C^1 on $(0, \eta)$, $\varphi'(s) > 0$ for all $s \in (0, \eta)$ and the KL inequality

$$\varphi'(\Psi(z) - \Psi(\overline{z}))\mathrm{dist}(0, \partial\Psi(z)) \geq 1. \tag{4}$$

holds for all $z \in U \cap [\Psi(\overline{z}) < \Psi < \Psi(\overline{z}) + \eta]$. If Ψ satisfies the KL property at each point of $\mathrm{dom}\partial\Psi$, then Ψ is called a KL function.
The convergence proof of SGP for KL functions follows the ideas presented in [2], in which the authors proved an abstract convergence result for descent methods satisfying the following three conditions:

Journal of Physics: Conference Series **756** (2016) 012001 doi:10.1088/1742-6596/756/1/012001

H1. (*Sufficient decrease condition*). There exists $a > 0$ such that $\Psi(x^{(k+1)}) + a\|x^{(k+1)} - x^{(k)}\|^2 \leq \Psi(x^{(k)})$ for each $k \in \mathbb{N}$.

H2. (*Relative error condition*). There exists $b > 0$ and, for each $k \in \mathbb{N}$, $w^{(k+1)} \in \partial\Psi(x^{(k+1)})$ such that $\|w^{(k+1)}\| \leq b\|x^{(k+1)} - x^{(k)}\|$.

H3. (*Continuity condition*). There exists a subsequence $\{x^{(k_j)}\}_{j\in\mathbb{N}}$ and $\tilde{x}$ such that $x^{(k_j)} \to \tilde{x}$ and $\Psi(x^{(k_j)}) \to \Psi(\tilde{x})$, as $j \to \infty$.

In our case, condition H3 is assured by the continuity of Ψ in Ω and the fact that $x^{(k)} \in \Omega$, for every $k \in \mathbb{N}$. Indeed H3 is needed in [2] only to ensure the stationarity of the limit point $\tilde{x}$, which has already been proved for SGP in [4].
Throughout the entire section, $\{x^{(k)}\}_{k\in\mathbb{N}}$ will denote the sequence generated by SGP. The following lemma (whose proof follows from [9, Lemma 4.1, Lemma 4.3 and Proposition 4.2]) and corollary guarantee that condition H1 holds for SGP.

Lemma 1 *The sequence* $\{x^{(k)}\}_{k\in\mathbb{N}}$ *satisfies*

$$\frac{1}{2\mu\alpha_{\max}}\|y^{(k)} - x^{(k)}\|^2 \leq -\nabla f(x^{(k)})^T(y^{(k)} - x^{(k)}). \tag{5}$$

Moreover, if ∇f *is L-Lipschitz continuous on* Ω*, then:*

- *there exists* $c > 0$ *such that for all* $\lambda \in [0, 1]$
$$f(x^{(k)} + \lambda d^{(k)}) \leq f(x^{(k)}) + \lambda(1 - cL\lambda)\nabla f(x^{(k)})^T d^{(k)}; \tag{6}$$
- *the sequence* $\{\lambda_k\}_{k\in\mathbb{N}}$ *of SGP linesearch parameters is bounded away from zero, i.e.,*
$$\lambda_k \geq \lambda_{\min}\ \forall k \in \mathbb{N}, \quad \lambda_{\min} > 0. \tag{7}$$
- *if* f *is also bounded from below, then* $0 < -\sum_{k=0}^{+\infty} \nabla f(x^{(k)})^T d^{(k)} < +\infty$ *and, consequently,*
$$\lim_{k\to+\infty} \nabla f(x^{(k)})^T d^{(k)} = 0. \tag{8}$$

Corollary 1 *If* ∇f *is L-Lipschitz continuous on* Ω*, then*

$$\Psi(x^{(k+1)}) + a\|x^{(k+1)} - x^{(k)}\|^2 \leq \Psi(x^{(k)}). \tag{H1}$$

Proof: Combining (5) with the backtracking rule (2) immediately yields

$$f(x^{(k+1)}) \leq f(x^{(k)}) - \frac{\beta\lambda_k}{2\alpha_{\max}\mu}\|y^{(k)} - x^{(k)}\|^2. \tag{9}$$

Because of (3), it is either $x^{(k+1)} = y^{(k)}$ or $x^{(k+1)} = x^{(k)} + \lambda_k d^{(k)}$. In the first case, using (9) with (7) leads to

$$f(x^{(k+1)}) \leq f(x^{(k)}) - \frac{\beta\lambda_{\min}}{2\alpha_{\max}\mu}\|x^{(k+1)} - x^{(k)}\|^2. \tag{10}$$

In the second case, we obtain the same inequality by using $y^{(k)} - x^{(k)} = \frac{1}{\lambda_k}(x^{(k+1)} - x^{(k)})$ and $\lambda_{\min} \leq \lambda_k \leq 1$ in (9). Finally, (H1) follows by adding the indicator function ι_Ω to both terms of (10) and by taking $a = \frac{\beta\lambda_{\min}}{2\alpha_{\max}\mu}$. □

We are now able to prove a slight variant of condition H2 for SGP, in which a subgradient of Ψ in $y^{(k)}$ (and not in $x^{(k)}$) is provided.

Lemma 2 *Suppose that* ∇f *is L-Lipschitz continuous on* Ω*. There exist* $b > 0$ *and* $v^{(k)} \in \partial\Psi(y^{(k)})$ *for all* $k \in \mathbb{N}$ *such that*

$$\|v^{(k)}\| \leq b\|x^{(k+1)} - x^{(k)}\|. \tag{H2}$$

Proof: By setting $\tilde{h}_k(y) = \nabla f(x^{(k)})^T(y - x^{(k)}) + \frac{1}{2\alpha_k}\|y - x^{(k)}\|^2_{D_k^{-1}} + \iota_\Omega(y)$, we rewrite $y^{(k)}$ as

$$y^{(k)} = \arg\min_{y \in \mathbb{R}^n} \tilde{h}_k(y) \Leftrightarrow 0 \in \partial \tilde{h}_k(y^{(k)}) \Leftrightarrow 0 \in \nabla f(x^{(k)}) + \frac{1}{\alpha_k} D_k^{-1}(y^{(k)} - x^{(k)}) + \partial \iota_\Omega(y^{(k)})$$
$$\Leftrightarrow w^{(k)} = -\nabla f(x^{(k)}) + \frac{1}{\alpha_k} D_k^{-1}(x^{(k)} - y^{(k)}) \in \partial \iota_\Omega(y^{(k)}).$$

Let us define $v^{(k)} = \nabla f(y^{(k)}) + w^{(k)} \in \partial \Psi(y^{(k)})$ for all $k \in \mathbb{N}$. By using the Lipschitz continuity of ∇f and (7), the following inequalities hold true:

$$\|v^{(k)}\| \leq \|\nabla f(y^{(k)}) - \nabla f(x^{(k)})\| + \frac{\mu}{\alpha_k}\|y^{(k)} - x^{(k)}\| \leq (L + \frac{\mu}{\alpha_{\min}})\|y^{(k)} - x^{(k)}\|$$
$$\leq \frac{1}{\lambda_k}(L + \frac{\mu}{\alpha_{\min}})\|x^{(k+1)} - x^{(k)}\| \leq \frac{1}{\lambda_{\min}}(L + \frac{\mu}{\alpha_{\min}})\|x^{(k+1)} - x^{(k)}\|.$$

The thesis holds by choosing $b = \frac{1}{\lambda_{\min}}(L + \frac{\mu}{\alpha_{\min}})$. □

3. Convergence result

In this section we present the convergence proof of SGP, which differs from that in [2, Lemma 2.6] by the presence of the projection step at each $k \in \mathbb{N}$, which was not considered in [2].

Theorem 1 *Suppose Ψ is a KL function and ∇f is L-Lipschitz continuous on Ω. If $\bar{x}$ is a limit point of $\{x^{(k)}\}_{k\in\mathbb{N}}$, then $\bar{x}$ is a stationary point for* (1) *and $x^{(k)}$ converges to $\bar{x}$.*

Proof: The stationarity of $\bar{x}$ has been proved in [4]. Since Ψ is a KL function, it satisfies the KL property at each point of Ω and, in particular, at $\bar{x}$. This means that there exist η, φ and U as in Section 2, such that the KL inequality (4) holds at $\bar{x}$. Equation (H1) implies that $\{\Psi(x^{(k)})\}_{k\in\mathbb{N}}$ is a non increasing sequence and hence $\Psi(x^{(k)}) \to \gamma$, $\Psi(x^{(k)}) \geq \gamma$. By the continuity of Ψ on Ω and the fact that $\bar{x}$ is a limit point of $\{x^{(k)}\}_{k\in\mathbb{N}}$, we deduce that $\gamma = \Psi(\bar{x})$. From (6) with $\lambda = 1$ and by definition (3) of $x^{(k+1)}$ we obtain

$$\Psi(x^{(k+1)}) \leq \Psi(y^{(k)}) \leq \Psi(x^{(k)}) + (1 - cL)\nabla f(x^{(k)})^T d^{(k)}. \quad (11)$$

Since $\Psi(x^{(k)}) \to \Psi(\bar{x})$ and (8) holds true, from (11) we also have that $\Psi(y^{(k)}) \to \Psi(\bar{x})$. Consequently, for all sufficiently large k we have

$$\Psi(\bar{x}) \leq \Psi(x^{(k)}) \leq \Psi(y^{(k-1)}) < \Psi(\bar{x}) + \eta. \quad (12)$$

Furthermore, let $\rho > 0$ be such that $B(\bar{x}, \rho) \subset U$. Then using the continuity of φ, one can choose $k_0 \in \mathbb{N}$ such that both (12) and the following technical condition are satisfied:

$$\|\bar{x} - x^{(k_0)}\| + 3\sqrt{\frac{\Psi(x^{(k_0)}) - \Psi(\bar{x})}{a\lambda^2_{\min}}} + \frac{b}{a}\varphi(\Psi(x^{(k_0)}) - \Psi(\bar{x})) < \rho. \quad (13)$$

We will now use $\{x^{(k)}\}_{k\in\mathbb{N}}$ to denote the sequence $\{x^{(k+k_0)}\}_{k\in\mathbb{N}}$. Let us rewrite (H1) as

$$\|x^{(k+1)} - x^{(k)}\| \leq \sqrt{\frac{\Psi(x^{(k)}) - \Psi(x^{(k+1)})}{a}}, \quad (14)$$

which, by recalling $x^{(k+1)} - x^{(k)} = \lambda_k(y^{(k)} - x^{(k)})$ and (7), yields also

$$\|y^{(k)} - x^{(k)}\| \leq \sqrt{\frac{\Psi(x^{(k)}) - \Psi(x^{(k+1)})}{a\lambda^2_{\min}}}. \quad (15)$$

Journal of Physics: Conference Series **756** (2016) 012001 doi:10.1088/1742-6596/756/1/012001

Fix $k \geq 1$. We state that if $x^{(k)}, y^{(k-1)} \in B(\bar{x}, \rho)$, then

$$2\|x^{(k+1)} - x^{(k)}\| \leq \|x^{(k)} - x^{(k-1)}\| + \varphi_k, \tag{16}$$

where $\varphi_k = \frac{b}{a}[\varphi(\Psi(x^{(k)}) - \Psi(\bar{x})) - \varphi(\Psi(x^{(k+1)}) - \Psi(\bar{x}))]$. Observe that, because of (12), the quantity $\varphi(\Psi(x^{(k)}) - \Psi(\bar{x}))$ makes sense for all $k \in \mathbb{N}$, and thus φ_k is well posed.
If $x^{(k+1)} = x^{(k)}$ inequality (16) holds trivially. Then we assume $x^{(k+1)} \neq x^{(k)}$. This assumption, combined with (14) and (12), guarantees that $x^{(k)}, y^{(k-1)} \in B(\bar{x}, \rho) \cap [\Psi(\bar{x}) < \Psi < \Psi(\bar{x}) + \eta]$ and therefore we can use the KL inequality in $x^{(k)}$ and $y^{(k-1)}$.
By exploiting the KL inequality at $y^{(k-1)}$ with (H2), it follows that $v^{(k-1)} \neq 0$ and $x^{(k-1)} \neq x^k$. Since $v^{(k-1)} \in \partial\Psi(y^{(k-1)})$, we can use again the KL inequality combined with (H2) to obtain

$$\varphi'(\Psi(y^{(k-1)}) - \Psi(\bar{x})) \geq \frac{1}{\|v^{(k-1)}\|} \geq \frac{1}{b\|x^{(k)} - x^{(k-1)}\|}. \tag{17}$$

Since φ is concave, its derivative is non increasing, thus $\Psi(y^{(k-1)}) - \Psi(\bar{x}) \geq \Psi(x^{(k)}) - \Psi(\bar{x})$ implies $\varphi'(\Psi(x^{(k)}) - \Psi(\bar{x})) \geq \varphi'(\Psi(y^{(k-1)}) - \Psi(\bar{x}))$. This fact applied to inequality (17) leads to

$$\varphi'(\Psi(x^{(k)}) - \Psi(\bar{x})) \geq \frac{1}{b\|x^{(k)} - x^{(k-1)}\|}. \tag{18}$$

Then following the same procedure of [2, Lemma 2.6], which uses the concavity of φ, (H1) and (18), we obtain $\|x^{(k+1)} - x^{(k)}\|^2 \leq \varphi_k\|x^{(k)} - x^{(k-1)}\|$ which, by applying the inequality $2\sqrt{uv} \leq u + v$, gives relation (16).
We are now going to show that for $j = 1, 2, \ldots$

$$x^{(j)}, y^{(j-1)} \in B(\bar{x}, \rho), \tag{19}$$

$$\sum_{i=1}^{j} \|x^{(i+1)} - x^{(i)}\| + \|x^{(j+1)} - x^{(j)}\| \leq \|x^{(1)} - x^{(0)}\| + \chi_j, \tag{20}$$

where $\chi_j = \frac{b}{a}[\varphi(\Psi(x^{(1)}) - \Psi(\bar{x})) - \varphi(\Psi(x^{(j+1)}) - \Psi(x))]$.
In order to prove (19)–(20), let us reason by induction on j. Using the triangle inequality, (14) with $k = 0$, the monotonicity of $\{\Psi(x^{(k)})\}_{k\in\mathbb{N}}$ and (13) we have

$$\|\bar{x} - x^{(1)}\| \leq \|\bar{x} - x^{(0)}\| + \|x^{(0)} - x^{(1)}\| \leq \|\bar{x} - x^{(0)}\| + \sqrt{\frac{\Psi(x^{(0)}) - \Psi(\bar{x})}{a}} < \rho,$$

namely $x^{(1)} \in B(\bar{x}, \rho)$. Using (15) with $k = 0$ and applying the same arguments as before, we also have $y^{(0)} \in B(\bar{x}, \rho)$. Finally, direct use of (16) shows that (20) holds with $j = 1$.
By induction, suppose that (19)–(20) hold for some $j \geq 1$. Since proving that $x^{(j+1)} \in B(\bar{x}, \rho)$ is identical to [2], we focus on $y^{(j)} \in B(\bar{x}, \rho)$. We rewrite (15) by noticing that $\Psi(\bar{x}) \leq \Psi(x^{(k+1)}) \leq \Psi(x^{(k)}) \leq \Psi(x^{(0)})$, which implies that $\|y^{(j)} - x^{(j)}\| \leq \sqrt{\left(\Psi(x^{(0)}) - \Psi(\bar{x})\right)/a\lambda_{\min}^2}$. By using this last equation, the triangle inequality, (20) with $k = j$ and (13), we have

$$\begin{aligned}
\|\bar{x} - y^{(j)}\| &\leq \|\bar{x} - x^{(0)}\| + \|x^{(0)} - x^{(1)}\| + \sum_{i=1}^{j} \|x^{(i+1)} - x^{(i)}\| + \|x^{(j+1)} - x^{(j)}\| + \|x^{(j)} - y^{(j)}\| \\
&\leq \|\bar{x} - x^{(0)}\| + 2\|x^{(0)} - x^{(1)}\| + \chi_j + \|x^{(j)} - y^{(j)}\| \\
&\leq \|\bar{x} - x^{(0)}\| + 3\sqrt{\frac{\Psi(x^{(0)}) - \Psi(\bar{x})}{a\lambda_{\min}^2}} + \frac{b}{a}\varphi(\Psi(x^{(0)}) - \Psi(\bar{x})) < \rho,
\end{aligned}$$

or equivalently $y^{(j)} \in B(\bar{x}, \rho)$. Hence, (16) holds with $k = j + 1$, i.e. $2\|x^{(j+2)} - x^{(j+1)}\| \leq \|x^{(j+1)} - x^{(j)}\| + \varphi_{j+1}$. Adding the above inequality with (20) (with $k = j$) yields (20) with $k = j + 1$, which completes the induction proof.
From (20) it immediately follows $\sum_{i=1}^{j} \|x^{(i+1)} - x^{(i)}\| \leq \|x^{(1)} - x^{(0)}\| + \frac{b}{a}\varphi(\Psi(x^{(1)}) - \Psi(\bar{x}))$ and therefore $\sum_{i=1}^{+\infty} \|x^{(i+1)} - x^{(i)}\| < +\infty$, which implies that the sequence $\{x^{(k)}\}_{k\in\mathbb{N}}$ converges to some x^*. Since $\bar{x}$ is a limit point of the sequence, it must be $x^* = \bar{x}$. □

Acknowledgments
This work has been partially supported by MIUR under the two projects FIRB - Futuro in Ricerca 2012, contract RBFR12M3AC and PRIN 2012, contract 2012MTE38N, and by the Italian GNCS - INdAM. I. Loris is a Research Associate of the Fonds de la Recherche Scientifique.

References

[1] H. Attouch, J. Bolte, P. Redont, and A. Soubeyran 2010, Proximal alternating minimization and projection methods for nonconvex problems: an approach based on the Kurdyka-Łojasiewicz inequality, *Math. Oper. Res.* **35**, 438–457.
[2] H. Attouch, J. Bolte, and B. F. Svaiter 2013, Convergence of descent methods for semi-algebraic and tame problems: proximal algorithms, forward-backward splitting, and regularized Gauss-Seidel methods, *Math. Program.* **137**, 91–129.
[3] J. Barzilai and J. M. Borwein 1988, Two-point step size gradient methods, *IMA J. Numer. Anal.* **8**, 141–148.
[4] S. Bonettini, R. Zanella, and L. Zanni 2009, A scaled gradient projection method for constrained image deblurring, *Inverse Probl.* **25**, 015002.
[5] S. Bonettini and M. Prato 2010, Nonnegative image reconstruction from sparse Fourier data: a new deconvolution algorithm, *Inverse Probl.* **26**, 095001.
[6] S. Bonettini, A. Cornelio, and M. Prato 2013, A new semiblind deconvolution approach for Fourier-based image restoration: an application in astronomy, *SIAM J. Imaging Sci.* **6**, 1736–1757.
[7] S. Bonettini and M. Prato 2014, Accelerated gradient methods for the X-ray imaging of solar flares, *Inverse Probl.* **30**, 055004.
[8] S. Bonettini and M. Prato 2015, New convergence results for the scaled gradient projection method, *Inverse Probl.* **31**, 095008.
[9] S. Bonettini, I. Loris, F. Porta, and M. Prato 2016, Variable metric inexact line–search based methods for nonsmooth optimization, *SIAM J. Optim.* **26**, 891–921.
[10] S. Bonettini, I. Loris, F. Porta, M. Prato, and S. Rebegoldi 2016, On the convergence of variable metric line-search based proximal-gradient method under the Kurdyka-Łojasiewicz inequality, *arxiv*:1605.03791.
[11] E. Chouzenoux, J.-C. Pesquet, and A. Repetti 2014, Variable metric forward–backward algorithm for minimizing the sum of a differentiable function and a convex function, *J. Optim. Theory Appl.* **162**, 107–132.
[12] P. L. Combettes and B. C. Vũ 2014, Variable metric forward-backward splitting with applications to monotone inclusions in duality, *Optimization* **63**, 17–31.
[13] R. Fletcher 2012, A limited memory steepest descent method, *Math. Program.* **135**, 413–436.
[14] K. Kurdyka 1998, On gradients of functions definable in o-minimal structures, *Ann. Inst. Fourier* **48**, 769–783.
[15] S. Łojasiewicz 1993, Sur la géométrie semi- et sous-analytique, *Ann. Inst. Fourier* **43**, 1575–1595.
[16] I. Loris, M. Bertero, C. De Mol, R. Zanella, and L. Zanni 2009, Accelerating gradient projection methods for ℓ_1-constrained signal recovery by steplength selection rules, *Appl. Computat. Harmon. A.* **27**, 247–254.
[17] M. Prato, R. Cavicchioli, L. Zanni, P. Boccacci, and M. Bertero 2012, Efficient deconvolution methods for astronomical imaging: algorithms and IDL-GPU codes, *Astron. Astrophys.* **539**, A133.
[18] M. Prato, A. La Camera, S. Bonettini, and M. Bertero 2013, A convergent blind deconvolution method for post-adaptive-optics astronomical imaging, *Inverse Probl.* **29**, 065017.
[19] Y. Yuan 2006, A new stepsize for the steepest descent method, *J. Comput. Math.* **24**, 149–156.
[20] B. Zhou, L. Gao, and Y. H. Dai: Gradient methods with adaptive step-sizes. *Comput. Optim. Appl.* **35**, 69–86.

Improving the accuracy of a Shack-Hartmann wavefront sensor on extended scenes

M Rais[1,2], J-M Morel[2], C Thiebaut[3], J-M Delvit[3] and G Facciolo[2,4]

[1] TAMI, Universitat des les Illes Balears, Spain
[2] CMLA, ENS-Cachan, Université Paris Saclay, France
[3] Centre National des Études Spatiales, France
[4] LIGM, UMR 8049, École des Ponts, UPE, France

E-mail: martin.rais@cmla.ens-cachan.fr

Abstract. In order to achieve higher resolutions, current earth-observation satellites use larger lightweight main mirrors which are usually deformed over time, impacting on image quality. In the context of active optics, we studied the problem of correcting this main mirror by performing wavefront estimation in a closed loop environment. To this end, a Shack-Hartman wavefront sensor (SHWFS) used on extended scenes could measure the incoming wavefront. The performance of the SHWFS on extended scenes depends entirely on the accuracy of the shift estimation algorithm employed, which should be fast enough to be executed on-board. In this paper we specifically deal with the problem of fast accurate shift estimation in this context. We propose a new algorithm, based on the global optical flow method, that estimates the shifts in linear time. In our experiments, our method proved to be more accurate and stable, as well as less sensitive to noise than all current state-of-the-art methods.

1. Introduction

Adaptive optics was originally developed for the field of astronomy to remove image aberrations induced by wavefronts propagating through Earth's atmosphere. Its task is to correct the aberrations of an incoming wavefront by using a deformable mirror in order to compensate for the distortion. It has been shown that adaptive optics allows to reduce these aberrations thus improving the image quality [1].

A key component of an adaptive optics system is the wavefront sensing mechanism. In astronomical imaging a wavefront sensing device is frequently used in conjunction with a deformable mirror in order to correct the undesired effects of atmospheric turbulence, thus improving the quality of sensed images. A Shack-Hartmann wavefront sensor (SHWFS) is one of such devices. It uses an array of lenslets to measure the deformation of the incoming wavefront. The shift of each lenslet focal plane image is proportional to the mean slope of the wavefront in the subaperture onto this lenslet, thus allowing to obtain a discrete local approximation of the slope of the wavefront, as shown in Fig. 1a. The measured slopes are then used to approximate the actual wavefront. This deformation is usually measured using point sources such as a star.

Adaptive optics can also be applied in the context of earth-observation satellites [2, 3]. In this setting, the problem of atmospheric turbulence is negligible. However, lightweight mirrors are required to increase the resolution, since mission costs are driven by the payload weight. The drawback of such mirrors is that time-varying deformations due to thermal effects and vibration

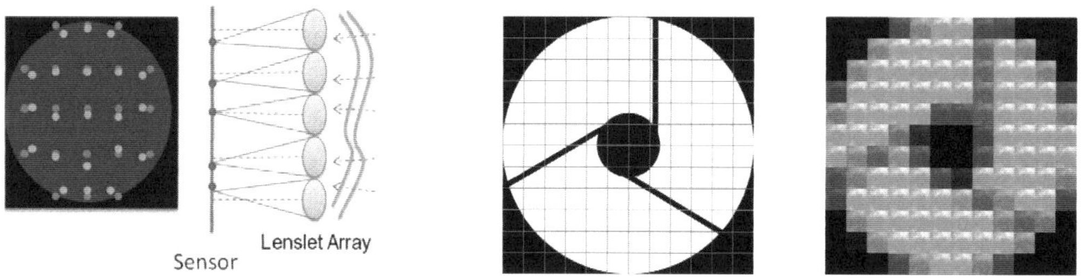

(a) Local shifts in the SHWFS (b) Occlusions in a SHWFS (c) Example images

Figure 1: Left: A SHWFS measures the wavefront by computing the local shifts between the detected spots (in red) and the reference spots (in grey), which would occur if no deformation is present. Right: A SHWFS occlusion schema using a Korsch telescope. The secondary mirror and the arms used to hold it are clearly visible. The decrease of the incoming signal for each lenslet is proportional to its occlusion.

deteriorate the image quality [4]. In adaptive optics a SHWFS device is used together with a deformable mirror to compensate for these aberrations.

There are some key differences between using wavefront sensing for earth-observation satellites and for astronomical observation. First, the SHWFS needs to be used on extended scenes, rather than source points. This yields a grid of images, one for each sub-aperture, which are shifted versions of the same scene. Second, the pupil of the Korsch telescope in the satellite is generally occluded in the center by a secondary mirror and the arms used to hold it. In these regions, the sub-apertures suffer a loss in the incoming signal, proportional to their percentage of occlusion (Fig. 1b), which results in the output depicted in Fig. 1c. Third, satellites have limited computational capacity which constrains the shift estimation task, making several shift estimation methods not suitable to be implemented on-board due to their high complexity.

Our contribution is threefold. First, an analysis of the *state-of-the-art* on wavefront correction through the use of SHWFS on extended scenes is performed (section 2). Second, we propose the use of an iterative global optical flow method for shift estimation (section 3) which presents several advantages over the conventional correlation methods. Finally, in section 4, a thorough evaluation of the discussed methods and our proposed algorithm is performed, evaluating their robustness to noise and image characteristics. Our conclusions are reported in section 5.

2. State-of-the-art review

Since the SHWFS was introduced in the late 1960s [5], several algorithms were proposed to estimate the shifts using point sources (like stars). However, only few authors have studied the problem when the source is extended, as when observing the earth from space. They can be organized in: Correlation methods which compute cross-correlation either in the space [6, 7, 8] or frequency domain [2, 9, 10, 11, 8] by assuming periodicity. Phase correlation methods estimate the shift directly in the frequency domain [11, 12] by fitting a plane to the phase correlation matrix between both images. The iterated methods [9, 10] progressively improve the estimation. Lastly, maximum likelihood (ML) approaches [13] incorporate specific noise models, and often lead to an optimization problem. Although it improves over correlation, ML is usually not considered for SHWFS [7, 2] because of its high computational cost.

Correlation-based Methods. These methods compute a correlation score on a grid $C(i, j)$ and interpolate it to determine the subpixel location of the peak. The methods mainly differ in the choices of correlation score and interpolation strategy. Michau et al. [6] was among the first to propose an experimental implementation for using a Shack-Hartmann sensor on extended sources. They compute the discrete cross-correlation of the two images, and estimate the subpixel shift by taking the centroid of the shifts with higher correlation value.

Löfdahl [8] tested several shift estimation algorithms considering several sources of errors such as noise, blur and bias mismatch. For the scores $C(i, j)$ he considered the sum of squared differences between the subimages (SDF), covariance in Fourier (CFI), a trend-corrected covariance (CFF) , the sum of the absolute differences (ADF), and its square (ADF^2). Subpixel shifts are computed by fitting a conic section around the minimum correlation value on the grid $C(i_{min}, j_{min})$, or by fitting parabolas independently in each dimension. For Gaussian noise, the authors choose SDF followed by fitting a 2D conic into the 3×3 values (SDF-2QI).

Poyneer [2] also evaluated several methods, including ML and a variant of phase correlation, but in the end opted for a similar Least Squares approach (using periodic-convolution computed in the Fourier domain) followed by 1D parabolic fitting.

Phase Correlation. Phase-correlation [12, 11] methods can estimate the shift directly in the frequency domain. In a noiseless situation, for an $M \times N$ image $f(x, y)$ the Fourier shift theorem yields $g(x, y) = f(x - \Delta_x, y - \Delta_y) = \mathcal{F}^{-1}(F(u, v)\exp(-j2\pi(\frac{u\Delta_x}{M} + \frac{v\Delta_y}{N})))$. Hence, the normalized cross-power spectrum between f and its shifted version g, given by $\frac{F(u,v)G^*(u,v)}{|F(u,v)G^*(u,v)|} = \exp(j2\pi(\frac{u\Delta_x}{M} + \frac{v\Delta_y}{N}))$, allows to compute the phase correlation matrix $\phi(u, v) = 2\pi(\frac{u\Delta_x}{M} + \frac{v\Delta_y}{N})$ for each frequency. The shifts can then be directly estimated in the frequency domain by fitting a plane passing through the origin of the computed correlation matrix [12]. To avoid aliased frequencies, Knutsson *et al.* [11] (also [9]) propose to discard most of the frequencies keeping only the two, or the four lowest frequencies, resulting in a fast method that does not require computing the whole FFT. The choice of two frequencies also has the advantage of being the least sensitive to aliasing [12], hence is the most reliable, but more sensitive to noise. Using more frequency components instead yields more robustness to noise.

Iterated Estimation. Sidick et al. [9] proposed ACC (Adaptive Cross-Correlation), an iterative method that estimates the shift in the Fourier domain by phase correlation [12]. Similarly to Knutsson et al. [11] only low frequencies are used to estimate the shift (just 8 coefficients). At each iteration, the second image is resampled (in the frequency domain) by the current shift estimate. This method attains very low errors of the order of 0.05 pixels by using fewer than 6 iterations. The same authors later [10] replaced the shift estimation with the periodic-correlation technique of [2] with parabolic fitting around the peak. The resulting method (APC) is more robust to noise than ACC.

3. Accurate shift estimation using optical flow

In order to estimate the shift between two sub-apertures, we propose to use a gradient-based shift estimator (GBSE) based on the optical flow equation. This technique, originally proposed by Lucas and Kanade [14], relates the difference between two successive frames to the spatial intensity gradient of the first image. Given the two images $I_2(x, y) = I_1(x - v_x, y - v_y)$ related by the translation (v_x, v_y), a first order Taylor approximation of I_1 yields the linear relation

$$I_1(x, y) - I_2(x, y) \approx v_x \frac{\partial I_1(x, y)}{\partial x} + v_y \frac{\partial I_1(x, y)}{\partial y}. \quad (1)$$

The displacement between I_1 and I_2 is estimated by computing the least squares solution

$$\begin{bmatrix} v_x \\ v_y \end{bmatrix} = \begin{bmatrix} \sum I_x^2 & \sum I_x I_y \\ \sum I_x I_y & \sum I_y^2 \end{bmatrix}^{-1} \begin{bmatrix} \sum I_x I_t \\ \sum I_y I_t \end{bmatrix}, \quad (2)$$

where I_x, I_y stands for $\frac{\partial I_1}{\partial x}$ and $\frac{\partial I_1}{\partial y}$ respectively, and $I_t = I_1 - I_2$ is the derivative over time.

It is important to remark that by centering the Taylor development at 0 and taking up to the first order term, the method gets systematically biased, directly affecting its accuracy. Furthermore, the noise in the input images is ignored by the algorithm, which also impacts on its performance. Both these bias sources are studied in detail in [15].

Iterative gradient-based shift estimator. Instead of dealing with the bias explicitly, Pham et al. [15] show that both bias sources depend linearly on the shift magnitude. This justifies the use of an iterative method, which is able to significantly reduce the bias, provided an appropriate resampling method is used. This algorithm is described in the following lines:

$i \leftarrow 0$; $I_2(0) \leftarrow I_2$; $w \leftarrow 0$
while $i < k$ **and** $|v(i-1) - v(i-2)| \geq \tau$ **do**
  $v(i) \leftarrow findshift(I_1, I_2(i))$ // Eq. (2)
  $w \leftarrow w + v(i)$ // Accumulate total shift
  if $i + 1 < k$ **then**
    $I_2(i+1) \leftarrow Shift(I_2, -w)$ // Resample the original I_2 by shifting it $-w$ pixels
  $i \leftarrow i + 1$

where k is the maximum amount of iterations, τ is a parameter that indicates the minimum incremental shift to stop iterating, $findshift$ uses Eq. (2) to solve for $v(i)$ and $Shift$ resamples an image by interpolation (see section 4).

Gradient computation. Several gradient estimation methods were tested and, due to the inherent noise in the input images, the Gaussian derivatives offered the best performance. Note that since our shift estimation is focused on accuracy and we are dealing with small images, using a larger kernel for computing the image derivatives imply discarding more boundary pixels, thus leaving fewer pixels to estimate the shift in Eq. (2).

Image intensities equalization. One requirement of the GBSE algorithm is to work on images with similar intensities. Since occluded sub-apertures receive less light, their intensities differ from the reference image. To this end, a simple intensity equalization applied. All subimages are equalized by scaling their mean to the highest mean among all subapertures, $I_k(x,y) = [\max_j(\mu_j)/\mu_k] I_k(x,y)$ where μ_k is the average value of image k.

4. Results

Experimental Setup. A simulator was used to evaluate the performance of the proposed method. Given a 12-bit input image, a set of images was generated matching a SHWFS configuration provided by CNES. A 12×12 lenslet grid occluded as shown in Fig. 1b was assumed. For each lenslet a ground truth shift was randomly generated. Since we were simulating in a closed loop environment, the maximum displacement was kept to half a pixel. Each sub-aperture image was then obtained by resampling the input image in Fourier, extracting a 37×37 pixels subimage, multiplying by the occlusion factor, and lastly adding the noise.

This setup permits compare SHWFS algorithms only considering the shifts induced by the wavefront slope. However, it has to be noted that due to phenomena such as scintillation and phase anisoplanatism, the relationship between the real wavefront aberration and the observed images may be more complex than shifts between the lenslet images.

The error for each method was computed as the mean error for all valid sub-apertures (less than 60% of its surface is occluded). The error for each sub-aperture is the Euclidean distance between the measured and the ground truth shift. To evaluate the robustness to image content, we used three subimages: a highly contrasted one that should not present any difficulty for shift estimation, a slightly more challenging one with its gradient mainly distributed on a single direction, and one from the sea with almost no signal. Different amounts of additive white Gaussian noise were simulated (with standard deviation $\sigma \in [1, ..., 150]$), and all methods were evaluated on 100 noise realizations. Fig. 2 shows the subimages affected by three noise levels. Signal dependent noise was handled by application of a variance stabilization transform.

We tested several variants of the proposed GBSE method. Five interpolation methods for resampling, namely bilinear, bicubic, cubic spline and Fourier (using the Fourier shift theorem), with and without image symmetrization, three values for the Gaussian derivative

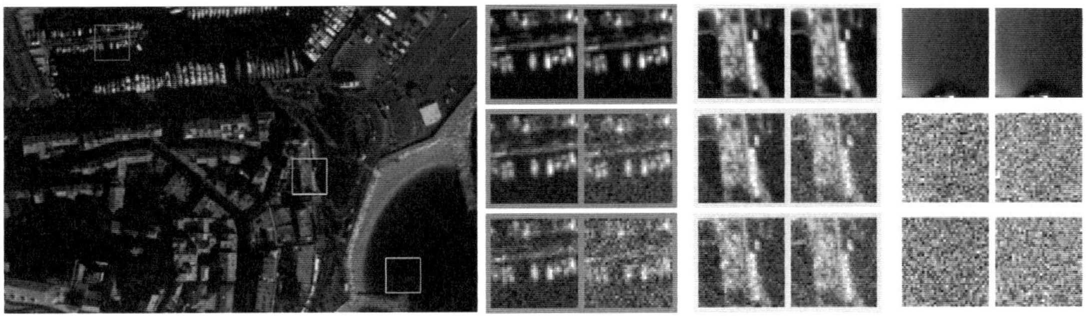

Figure 2: Left: Input image used for the simulation. Right: For each level of noise $\sigma = \{1, 100, 200\}$ (vertically separated), two different sub-apertures are shown: no occlusion and 57% occluded (horizontally separated). The dynamic ranges were stretched for viewing purposes.

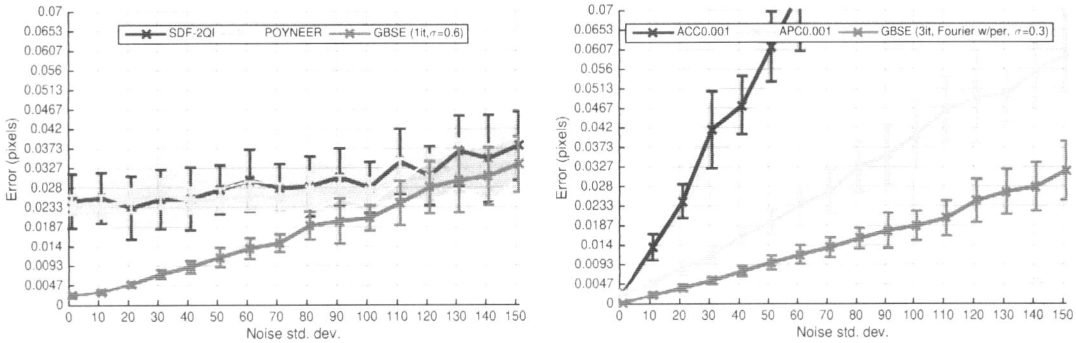

Figure 3: Performance for non-iterative and iterative methods on first test image.

width $\sigma = \{0.3, 0.6, 1\}$, and using up to three iterations per shift estimation. From these methods we retained the best performing non-iterative method ($k = 1$ and no resampling) which uses $\sigma = 0.6$ for the gradient computation, and the best with 3 iterations which uses Fourier resampling with image periodization and $\sigma = 0.3$. We compared them with the best performing state-of-the-art algorithms, namely, ACC [9], APC [10], the method from Poyneer [2] and the SDF-2QI algorithm from Löfdahl [8].

It should be noted that if a single iteration is used, the gradient estimation with $\sigma = 0.6$ gives better results than using $\sigma = 0.3$. This is because a higher σ removes more noise from the derivatives, and blurring the images impacts less on the accuracy than not taking care of the noise. This noise influence can be reduced by iterating the method [15], which justifies estimating the gradient from a sharper image on those cases.

Results and discussion. We compared iterative and non-iterative algorithms separately. The average errors for non-iterating methods on the first image type are shown in Fig. 3 (left). The non-iterative GBSE outperforms both Poyneer and SDF-2QI (which perform similarly in this image), GBSE is also more stable (less variability). In Fig. 3 (right) both iterative methods ACC and APC were compared with our GBSE using 3 iterations. Again the proposed method is the most accurate and more stable. Even more, this method is the best also compared to non-iterated methods. We also note that ACC is not much robust to noise. This is a well-known problem for phase-correlation methods which use a limited number of frequencies [16].

The results for the second image type (Fig. 4) are similar to the first one, although the average errors are higher. While the SDF-2QI performs slightly worse, the Poyneer method is considerably worse. This is due to discrepancies at the image boundaries resulting from the periodicity assumption of the periodic convolution used by Poyneer. This is also explains why the ACC method behaves so poorly on this image. Yet again, the proposed methods are more precise, stable and with less variability than the state-of-the-art. Finally, for the third image

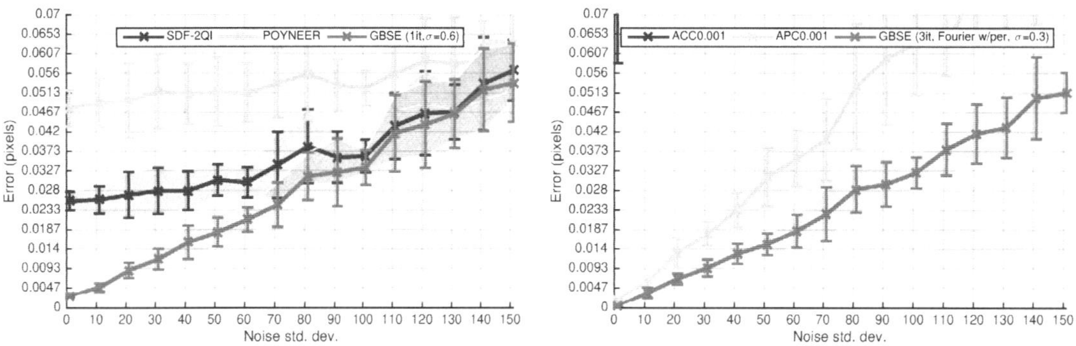

Figure 4: Performance for non-iterative and iterative methods on second test image.

type all the methods fail so the graph is omitted.

5. Conclusion

A new method for accurate sub-pixel shift-estimation, based on an iterative global optical flow algorithm, is proposed in the context of a SHWFS on extended scenes for aberration correction on-board earth-observation satellites. Simulations using plausible information provided by CNES show that the proposed algorithm is more accurate, stable, robust to noise, and with lower variability than the current state-of-the-art, permitting a more precise wavefront estimation.

Acknowledgments

Work partly founded by the Centre National d'Etudes Spatiales (CNES, MISS Project) and the Office of Naval research (ONR grant N00014-14-1-0023). During this work, Martin Rais had a fellowship of the Ministerio de Economia y Competividad (Spain), reference BES-2012-057113, for the realization of his Ph.D. thesis.

References

[1] Wizinowich P *et al.* 2000 *Publications of the Astronomical Society of the Pacific* **112** 315–319
[2] Poyneer L A 2003 *Applied Optics* **42** 5807–5815
[3] Bonnefois A *et al.* 2014 Comparative theoretical and experimental study of a Shack-Hartmann and a Phase Diversity Sensor, for high-precision wavefront sensing dedicated to Space Active Optics *ICSO 2014*
[4] Chen P C, Bowers C W, Marzouk M and Romeo R C 2000 *Optical Engineering* **39** 2320–2329
[5] Platt B C *et al.* 2001 *Journal of Refractive Surgery* **17** S573–S577
[6] Michau V, Rousset G and Fontanella J 1993 Wavefront sensing from extended sources *Real Time and Post Facto Solar Image Correction* vol 1 p 124
[7] Michau V, Conan J M, Fusco T, Nicolle M, Robert C, Velluet M T and Piganeau E 2006 Shack-hartmann wavefront sensing with extended sources *SPIE Optics+Photonics* (SPIE) pp 63030B–63030B
[8] Löfdahl M G 2010 *Astronomy & Astrophysics* **524** A90
[9] Sidick E *et al.* 2007 An Adaptive cross-correlation algorithm for extended scene Shack-Hartmann wavefront sensing *Adaptive Optics: Methods, Analysis and Applications* (OSA)
[10] Sidick E 2011 Adaptive periodic-correlation algorithm for extended scene shack-hartmann wavefront sensing *Imaging and Applied Optics* (Optical Society of America) p CPDP1
[11] Knutsson P, Owner-Petersen M and Dainty C 2005 *Optics express* **13** 9527–9536 ISSN 1094-4087
[12] Stone H, Orchard M, Chang E C and Martucci S 2001 *Geoscience and Remote Sensing, IEEE Transactions on* **39** 2235–2243 ISSN 0196-2892
[13] Gratadour D, Mugnier L and Rouan D 2005 *Astronomy & Astrophysics* **443** 357–365
[14] Lucas B D and Kanade T 1981 An iterative image registration technique with an application to stereo vision IJCAI'81 (Morgan Kaufmann Publishers Inc.) pp 674–679
[15] Pham T Q, Bezuijen M, Van Vliet L J, Schutte K and Hendriks C L L 2005 Performance of optimal registration estimators *Proc. SPIE* vol 5817 pp 133–144
[16] Gilman A and Leist A 2013 Global Illumination-Invariant Fast Sub-Pixel Image Registration pp 95–100 ISBN 978-1-61208-283-7

Contrast enhanced computerized tomography measurement of vascular blood flow

B Sixou[1], **L.Boissel**[1], **M.Sigovan**[1]

[1] CREATIS, INSA-Lyon; Inserm, U1044; CNRS UMR 5220; Université Lyon 1; Université de Lyon; 69621, Lyon, France

E-mail: `bruno.sixou@insa-lyon.fr`

Abstract. In this work, we study the measurement of blood velocity with contrast enhanced computed tomography. The transport equation is used as a constraint to obtain stable solutions. The inverse problem is formulated as an optimal control problem. The density of the contrast agent is reconstructed together with the flow field. The inversion scheme is tested on a simple phantom. The reconstruction of the velocity is improved but the convergence of the method is slow.

1. Introduction

The blood velocity in arteries is an important clinical parameter. It can be evaluated with contrast agent enhanced tomography. Yet, in contrast enhanced computed tomography imaging this velocity measurement is not a well established method. Some blood velocity measurement with CT is known as the time-of-flight TOF technique and is based on the difference in arrival time of the contrast agent at two different locations. This measurement of the intravascular blood velocity is based on the temporal changes in the sinogram obtained in sequential scanning mode. The method is limited to the estimation of the mean of the speed of the flow component along the flow field propagation axis [1].

The reconstruction of the flow velocity and of the density of the contrast agent can be formulated as a dynamic inverse problem in which the measuring process and the data are time dependent [2, 4, 10]. Such problems arise in the field of medical imaging for moving or beating organs. In order to obtain stable solutions, time-dependent regularization techniques are required [5, 6, 7, 8]. Some regularization methods based on optimal control problem can be found in [2].

We present here a method to evaluate the blood velocity from spiral CT scans, consisting of repeated spiral acquisitions with alternating scan directions. The method is based on the integration of physical constraints on the flow for the iterative reconstruction. The scalar transport equation and the divergence free condition are used for the reconstruction of two types of functions: the density of the contrast agent $f(\mathbf{x}, t)$ and the flow field $\mathbf{w}(\mathbf{x}, t)$ for $x \in \mathbb{R}^3$ and for $x \in [0, T]$. The inverse problem is thus adressed in a 3D+t framework. The main contribution of this work is to formulate the blood velocity and density reconstruction problem as an optimal control problem. We consider here a simple geometry where the flow field to recover is uniform to validate the approach.

This paper is organized as follows. In Section 1, the dynamic tomography problem considered is presented and it is formulated as an optimal control problem. To solve this problem, we derive first-order optimality conditions. We obtain a system of PDE with initial and boundary conditions. We display numerical results with simple synthetic images that demonstrate the efficiency of the optimal control formulation to improve the reconstruction of the flow field and of the density of the contrast agent.

2. Optimal control formulation for dynamic tomography

The flow phantom and the CT acquisition geometry used in this study are displayed in Figure 1. This phantom simulates a straight arterial vessel segment oriented along the z axis. We consider here a 4-dimensional (4D) spiral acquisition mode, consisting of repeated spiral acquisitions with projection directions perpendicular to the z axis of rotation.

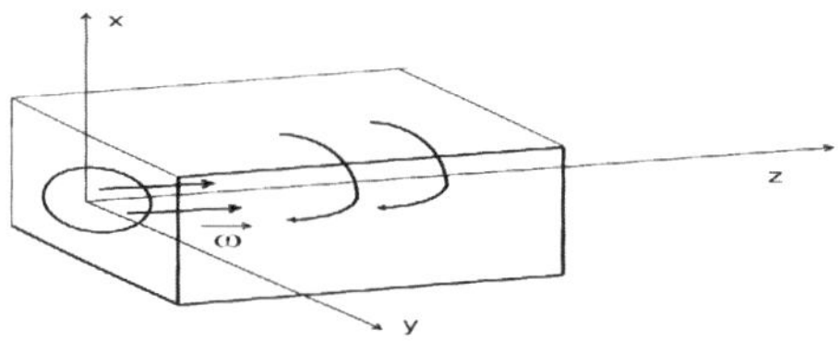

Figure 1. Flow phantom and CT acquisition geometry.

.

We first formulate the optimal control problem with spatio-temporal regularization in an infinite dimensional setting. The spatial domain is a bounded open domain Ω. The density of the contrast agent to be reconstructed is denoted as $f(\mathbf{x}, t)$. The space-time box for the problem is denoted as $Q = \Omega \times [0, T]$, where $[0, T]$ is the time domain. We are only interested in the contrast agent and the anatomical structures other than the contrast material are assumed to have a zero density. For each time $t \in [0, T]$, a projection angle $\theta(t)$ is used. Let $X = L_2(\Omega)$, our aim is to find $f : [0, T] \to X$ from the linear equations:

$$R_{\theta(t)} f(\mathbf{x}, t) = p_{\theta(t)} \quad \forall t \in [0, T] \tag{1}$$

where $p_{\theta(t)}$ are the dynamic measured projections, $R_{\theta(t)}$ is the linear Radon operator for the projection angles $\theta(t)$ that is available at time t.

We assume the contrast agent follows a scalar transport equation with a velocity field, $\mathbf{w} = (u, v, w) : [0, T] \times \Omega \to \mathbb{R}^3$. It is injected on a disk D in the plane $z = 0$. The following transport equation is satisfied:

$$\left\{ \begin{array}{c} E(f) = \frac{\partial f(\mathbf{x},t)}{\partial t} + u\frac{\partial f(\mathbf{x},t)}{\partial x} + v\frac{\partial f(\mathbf{x},t)}{\partial y} + w\frac{\partial f(\mathbf{x},t)}{\partial z} = 0 \quad \forall t \in [0, T] \quad \forall \mathbf{x} \in \Omega \\ f(\mathbf{x}, t) = 1 \;\; on \;\; D \;\; \forall t \in [0, T] \\ f(\mathbf{x}, 0) = 0 \;\; on \;\; \Omega - D \end{array} \right. \tag{2}$$

We have added a divergence-free constraint of the flow field $\mathbf{w}$, $\text{div}(\mathbf{w}) = 0$ to make the flow smooth and varying not too much inside the moving object. This transport equation and the divergence free condition will be used as a constraints to obtain stable solutions. The cost functional without any constraint can be written:

$$J(f,\mathbf{w}) = \ \tfrac{1}{2}\int_0^T \|R_{\theta(t)}f(\mathbf{x},t) - p^{\delta}_{\theta(t)}(t)\|^2 dt + \tfrac{\alpha}{2}\int_0^T \|\nabla\mathbf{w}\|_2^2 dt \tag{3}$$

where $p^{\delta}_{\theta(t)}$ represent the noisy projections and $\|\nabla\mathbf{w}\|_2$ the Frobenius norm of the Jacobian matrix of the flow field $\mathbf{w}$. The first term is a data term that measures the discrepancy between the measured projections and the projected images. The term $\frac{\alpha}{2}\int_0^T \|\nabla\mathbf{w}\|_2^2 dt$ is a smoothing term and α a regularization parameter. We consider two dual variables $p : [0,T] \times \Omega \to \mathbb{R}$ and $q : [0,T] \times \Omega \to \mathbb{R}$ for the constraints, and the Lagrangian L of the constrained problem is given by;

$$L(f,\mathbf{w},p,q) = J(f,\mathbf{w}) + \beta\int_0^T < Ef(t), p(t) > dt + \gamma\int_0^T < \nabla.\mathbf{w}), q(t) > dt \tag{4}$$

where β and γ are regularization parameters.

2.1. Optimality system

To minimize the functional, we have to derive the first order optimality condtion with the Euler-Lagrange equations, from which the optimal state f, the optimal speed $\mathbf{w}$, and the adjoint variable p and q can be determined.

2.1.1. Optimality system of p The adjoint equation specifies the first-order necessay condition with respect to the state variable f. The first variation of the Lagrangian L at $f \in L_2(Q)$ in the direction of $h \in L_2(Q)$ is defined by [11]:

$$\delta L(f,\mathbf{w},p,q)(h) = \frac{\partial L(f+\epsilon h,\mathbf{w},p,q)(h)}{\partial\epsilon}|_{\epsilon=0} \tag{5}$$

If a minimizer of L exists, then the equation $\delta L(f,\mathbf{w},p,q)(h) = 0$ must hold for every $h \in L_2(Q)$.

$$\delta L(f,\mathbf{w},p,q)(h) = \beta\int_0^T < Eh(t), \mu(t) > dt + \int_0^T < R_{\theta(t)}h, R_{\theta(t)}f - g > dt \tag{6}$$

$$\int_0^T < R_{\theta(t)}h, R_{\theta(t)}f - g > dt = \int_0^T < h, R^t_{\theta(t)}(R_{\theta(t)}f - g) > dt \tag{7}$$

The second term can be written:

$$\int_0^T < Eh(t), p(t) > dt = \int_0^T < \frac{\partial h}{\partial t} + u\frac{\partial h}{\partial x} + v\frac{\partial h}{\partial y} + w\frac{\partial h}{\partial z}, p(t) > dt \tag{8}$$

With integration by parts,

$$\int_0^T < \frac{\partial h}{\partial t}, p(t) > dt = -\int_0^T < \frac{dp}{dt}, h(t) > dt + < h(t), p(t) > |_0^T = -\int_0^T < \frac{dp}{dt}, h(t) > dt \tag{9}$$

with $p(T) = 0$. The second term can be written:

$$\int_0^T < p\mathbf{w}, grad(h) > dt = -\int_0^T div(p\mathbf{w})h dt \tag{10}$$

With the divergence free condition, $\mathrm{div}(\mathbf{w}) = 0$, and $div(p\mathbf{w}) = p div(\mathbf{w}) + grad(p).\mathbf{w} = grad(p).\mathbf{w}$. We obtain the following equation for p:

$$\beta(\frac{dp}{dt} + grad(p).\mathbf{w}) = R^t_{\theta(t)}(R_{\theta(t)}f - g) \ \ p(.,T) = 0 \tag{11}$$

This adjoint equation is a backward equation.

2.1.2. Optimality system for the speed For the optimality condition for the speed, we have to write, $\delta L(f, \mathbf{w}, p, q)(\mathbf{h}) = 0$, $\forall \mathbf{h} \in L_2(Q)^3$. We obtain:

$$\alpha \int_0^T < \nabla \mathbf{w}, \nabla \mathbf{h} > dt + \gamma \int_0^T < \nabla.\mathbf{h}, q(t) > dt + \beta \int_0^T < p\mathbf{h}, \nabla(f) > dt = 0 \quad (12)$$

Since

$$\alpha \int_0^T < \nabla \mathbf{w}, \nabla \mathbf{h} > dt = -\alpha \int_0^T \triangle \mathbf{w} h dt \quad (13)$$

$$\gamma \int_0^T < \nabla.\mathbf{h}, q(t) > dt = -\gamma \int_0^T \mathbf{h}.\nabla(q) dt \quad (14)$$

$$\beta \int_0^T < p\mathbf{h}, \nabla(f) > dt = \beta \int_0^T < \mathbf{h}, p\nabla(f) > dt \quad (15)$$

we obtain, with the variation with respect to the speed:

$$\alpha \triangle \mathbf{w} + \gamma \nabla q = \beta p \nabla f \quad in \quad \Omega \times [0, T] \quad (16)$$

3. Numerical algorithm

3.1. Simulation details

The values of the components (u, v, w) of the flow field $\mathbf{w}$ and of the density f are discretized on a grid of size $N_s = 53 \times 53 \times 53$ for the spatial domain Ω and on a grid of size $N_t = 360$ for the time domain $[0, T]$. For each time t_k in the time domain $[0, T]$, a projection angle $\theta_k = 2\pi t_k / T$ is used for the Radon projection operator. During the CT gantry rotation, 360 parallel projections are thus acquired. Moreover, 77 projections rays are acquired for each value of the projection angle. The ground truth solution f^* is obtained from the scalar transport equation with a uniform velocity along the z axis, $\mathbf{w}^* = (u^*, v^*, w^*) = (0, 0, 15)$. The initial density of the contrast agent is $f(\mathbf{x}, z = 0, t = 0) = 1$ for the pixels inside a disk of diameter 17 around the z axis. A Gaussian white noise with peak-to-peak signal-to-noise ratio (PPSNR) of 48 dB was added to the projections. Our algorithm is not globally convergent algorithms. The scalar diffusion transport equation (Eq.2) was then used to obtain a first guess of the density profile $f_0(\mathbf{x}, t)$. This a priori guess of the solution ensures the convergence of the algorithms.

The following loop was used for the numerical solution the optimal control problem:
1) Solve the forward flow equation to obtain $f_k(\mathbf{x}, t)$ from the current approximation $\mathbf{w}_k$(Eq.2). For the integration of the partial differential equation, the problem is discretized by finite difference and we use an explicit scheme of order 2 with numerical diffusion. For this numerical scheme, the CFL condition is verified, $CFL = \|\mathbf{w}\| dt/dx < 1$. For the simulation of the transport equation, the time step $dt = 0.005$ is used.
2) Solve the backward adjoint equation to obtain the adjoint variable p_k(Eq.12). The adjoint equation is solved by the same method as described before by reversing time.
3) Update the flow field $\mathbf{w}_k$. Following the work of Tai et al. [9], for the TV-Stokes equation, rather than solving the Eq.17 directly, we use a "gradient descent" strategy to determine the flow $\mathbf{w}_k$. Given an initial estimate $\mathbf{w}_k^0 = \mathbf{w}_{k-1}$, the gradient flow equation for $\mathbf{w}_k$ is given by:

$$\frac{\partial \mathbf{w}_k}{\partial t} = \alpha \triangle \mathbf{w}_k + \gamma \nabla q_k - \beta p_k \nabla f_k \quad in \quad \Omega \times [0, T] \quad (17)$$

The search for a minimizer is done using the following iterative procedure to update $\mathbf{w}_k$ and q_k with time step $\triangle t$:

$$\mathbf{w}_k^{n+1}(t) = \mathbf{w}_k^n(t) + \triangle t(\alpha \triangle \mathbf{w}_k^n(t) + \gamma \nabla q_k^n(t) - \beta p_k(t) \nabla f_k(t)) \quad (18)$$

$$q_k^{n+1}(t) = q_k^n(t) + \triangle t \nabla . \mathbf{w}_k^n(t) \tag{19}$$

The asymptotic state of w_k when $n \to \infty$ is chosen as the solution of the Eq.18. $\triangle t$ is chosen very small to have a stable scheme. With large $\triangle t$, the iterations will converge faster but if they are two large, the numerical scheme is unstable. For our numerical experiments, the value $\triangle t = 0.005$ leads to convergence [9]. The regularization parameters α, β and γ are chosen to have the best decrease of the discrepancy term. The values α, β, γ used in this work are 2.10^{-1}, 2.10^{-2} and 2.10^{-2} respectively.

3.2. Results

In order to validate the method, we have calculated the error term for the velocity field and the reconstructed density with L_2 norms

$$D(f_k) = \sum_{k=1}^{N} \|f_k(\mathbf{x}, t_k) - f^*(\mathbf{x}, t_k)\|^2 \tag{20}$$

$$D(\mathbf{w}_k) = \sum_{k=1}^{N} \|u^* - u_k\|_2^2 + \|v^* - v_k\|_2^2 + \|w^* - w_k\|_2^2 \tag{21}$$

Sections of the true and reconstructed w maps obtained at the end of the iterations for $z = 10$ and $t = 100s$ are displayed in Figure 2. Sections of the true and reconstructed contrast agent density maps obtained at the end of the iterations for $z = 10$ and $t = 100s$ are displayed in Figure 3. Some plots of the error decreases are displayed in Figure 4 for the discrepancy terms $D(f)$ and $D(\mathbf{w})$. A large reduction of the error is obtained for the reconstructed density. The convergence speed for the velocity field is slower.

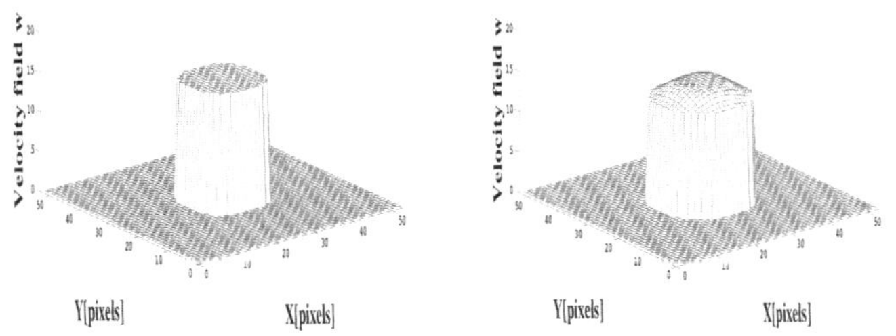

Figure 2. Sections of the true and reconstructed w maps for z=10 and t=100.

4. Conclusion

In this work, we consider the CT measurement of contrast agent flow velocity by using projection data obtained during the spiral acquisition of projections data around the sample as the contrast material flows into the phantom. An optimal formulation of the problem is used in which the scalar diffusion equaition is used. The optimality system for the density of the contrast agent and the flow field are solved iteratively. The efficiency of the method is shown wit a dynamic flow phantom. The reconstruction eror for the contrast agent density is much decreased but the convergence of the velocity field is slow.

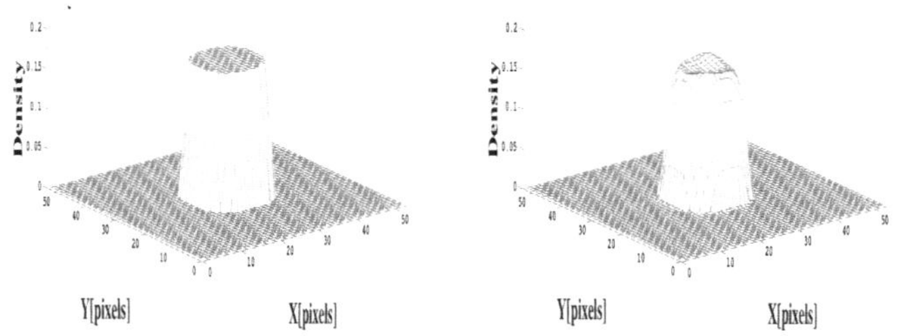

Figure 3. Sections of the true and reconstructed contrast agent density maps for z=10 and t=100.

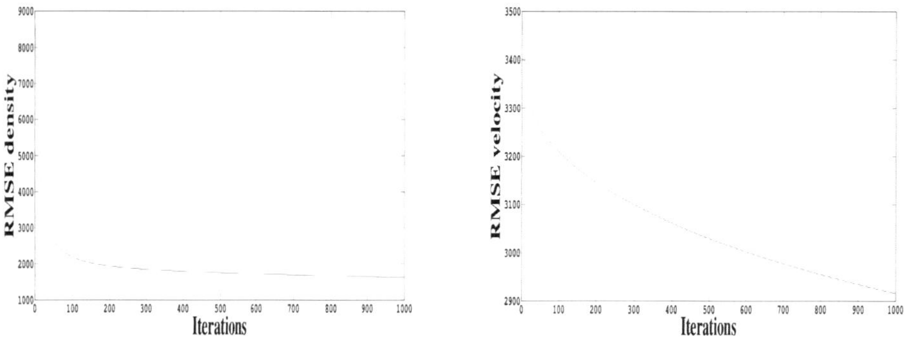

Figure 4. Evolution of error for the flow velocity and the density of the contrast agent.

References

[1] S.Prevrhal, C.H.Forsythe, R.J.Harnish, M.Saeed, B.M.Yeh "CT angiographic measurement of vascular blood fowl velocity by using projection data", *Radiology*, vol.261, 923-929 , 2011.

[2] S.Kinderman and A.Leitao "On regularization methods for inverse problems of dynamic type", *Numerical Functional Analysis and Optimization*, vol.27, 139-160 , 2006.

[3] A.Borzi, K.Ito and K.Kunisch "Optimal control approach to optical flow computation", *International Journal for Numerical Methods in Fluids*, vol.40, 231-240 , 2002.

[4] C.Mory, V.Auvray, B/Zhang, M.Grass, D.Schafer, S.J.Chen, J.D.Carrol, S.Rit, F.Peyrin, P.Douek and L.Boussel "Cardiac C-arm computed tomography using a 3d+time ROI reconstruction method with spatial and temporal regularization", *Medical Physics* , vol.41, 021903 , 2014.

[5] H.Ammari, G.Bao and J..Fleming "An inverse source problem for Maxwell'equations in magnetoencephatograhy", *SIAM Journal of Applied Mathematics*, vol.62, 1369-1382 , 2002.

[6] U.Schmitt, A.K.Louis,C.Wolters and M.Vauhkonen "Efficient algorithms for the regularization of dynamic inverse problems I:Theory", *Inverse Problems*, vol.18, 645-658 , 2002.

[7] U.Schmitt, A.K.Louis,C.Wolters and M.Vauhkonen "Efficient algorithms for the regularization of dynamic inverse problems II:Applications", *Inverse Problems*, vol.18, 659-676 , 2002.

[8] A.Seppnen, M.Vauhkonen, P.J.Vauhkonen, E.Sommersalo and J.P.Kaipio "State estimation with fluid dynamical evolution models in process tomography-an application to impedance tomography", *Inverse Problems*, vol.17, 467-483 , 2001.

[9] X.X.Tai, S.Osher and R.Holm "Image impainting using a TV-Stokes equation", *Image Processing based on partial differential equations*, 3-22, Springer Berlin Heidelberg, 2007.

[10] F.Momey, E.Thiebaut, C.Burnier, L.Denis, J.M.Becker and L.Desbat. "Regularized 4D-CT reconstruction form a single data set with a spatio-temporal prior.", hal-00998291v1, 2015 .

[11] R.Courant *Calcul of Variations*, New York, 1946.

Journal of Physics: Conference Series **756** (2016) 012004 doi:10.1088/1742-6596/756/1/012004

Efficient Compressed Sensing Based MRI Reconstruction using Nonconvex Total Variation Penalties

D. Lazzaro, E. Loli Piccolomini, F. Zama

Department of Mathematics, University of Bologna

E-mail: fabiana.zama@unibo.it

Abstract. This work addresses the problem of Magnetic Resonance Image Reconstruction from highly sub-sampled measurements in the Fourier domain. It is modeled as a constrained minimization problem, where the objective function is a non-convex function of the gradient of the unknown image and the constraints are given by the data fidelity term. We propose an algorithm, Fast Non Convex Reweighted (FNCR), where the constrained problem is solved by a reweighting scheme, as a strategy to overcome the non-convexity of the objective function, with an adaptive adjustment of the penalization parameter. We propose a fast iterative algorithm and we can prove that it converges to a local minimum because the constrained problem satisfies the Kurdyka-Lojasiewicz property. Moreover the adaptation of non convex l0 approximation and penalization parameters, by means of a continuation technique, allows us to obtain good quality solutions, avoiding to get stuck in unwanted local minima. Some numerical experiments performed on MRI sub-sampled data show the efficiency of the algorithm and the accuracy of the solution.

1. Introduction

Compressed Sensing (CS) is based on three major components: the sparsity of representation in some known domain, the incoherent measurements, which are usually achieved by irregular sampling patterns such as random sampling, and finally the sparsity-constrained non-linear reconstruction. Thanks to these properties CS exploits the compressibility of medical images to reconstruct data sampled below the Nyquist rate (undersampled), without loss of information. The technique can be applied to Magnetic Resonance Imaging (MRI) to increase imaging speed or to Computed Tomography (CT) to reduce radiation dose. MRI is a particular good candidate for the application of compressed sensing since data acquisition is performed in a transform domain (Fourier space) instead of the image domain, which facilitates the generation of incoherent aliasing artifacts. Many MRI CS imaging applications are modeled as constrained l_1 minimization problems:

$$\min_u \|\Psi(u)\|_1 \quad s.t. \quad \Phi u = z \tag{1}$$

where Ψ is a sparsity promoting operator, such as wavelet or discrete gradient, z are the undersampled data in the Fourier domain, $\Phi \in \mathbb{R}^{M\times N}, M < N$ is the data acquisition matrix and u is the reconstructed image. However the number of samples required to obtain good reconstructions is much larger compared to the minimum sampling rate that can be obtained

Journal of Physics: Conference Series **756** (2016) 012004 doi:10.1088/1742-6596/756/1/012004

by l_0 minimization. For this reason, some authors (see for example [1], [2]) consider non-convex approximations of the l_0 norm, in order to reduce the sampling in the Fourier space. Following this approach, we define the objective function in (1) by approximating the l_0 norm with a non-convex function of the image gradient ($F(Du)$), and we consider the Lagrangian formulation:

$$\min_u \left\{ F(Du) + \frac{1}{2\lambda}||\Phi u - z||_2^2 \right\}, \quad \lambda > 0 \tag{2}$$

where λ is the regularization parameter, balancing the least squares data fidelity term and the non convex penalty term. One of the most outstanding technique for the solution of the non-convex minimization problem (2) is the Iterative Reweighted (IR) ℓ_1 scheme [3] where a local linear approximation of the non-convex $F(Du)$ is considered. Hence the problem (2) is approximated by a sequence of convex minimization problems. Each convex minimization problem that we obtain is solved by means of a forward-backward splitting algorithm, where the backward step, usually the most computationally expensive, is solved by means of the Weighted Split Bregman method [4], modified for the anisotropic case and suitably accelerated. This allows us to drastically reduce the computational costs.

The paper is organized as follows. In section 2 we state the problem to be solved and we describe the proposed algorithm; in section 3 some numerical experiments on MRI undersampled synthetic data are presented and finally in section 4 we report our conclusions.

2. The method

In this section we give a precise definition of the objective function $F(Du)$ in (2) and, after analyzing its properties we formalize our algorithm. We define $F(Du)$ as follows:

$$F(Du) = \sum_{i,j} (\psi_\mu(|u_x|_{i,j}) + \psi_\mu(|u_y|_{i,j}))$$

where $Du \in \mathbb{R}^{N \times 2N}$ is the gradient of the image u, whose components (u_x, u_y) are obtained by backward finite differences approximations of the partial derivatives in the x and y spatial coordinates respectively. The function $\psi : \mathbb{R} \to \mathbb{R}$ is a continuous, symmetric, twice differentiable on $\mathbb{R} \setminus \{0\}$, increasing and non-convex function, whose goal is to improve the sparsity promoting properties of ℓ_1 norm. Among the different examples reported the literature [3, 5], in this work we choose the following function:

$$\psi_\mu(t) = \frac{1}{\log(2)} \log\left(\frac{2}{1 + e^{-\frac{|t|}{\mu}}}\right), \quad \mu > 0$$

where μ is a positive constant such that, as $\mu \to 0$, the proposed $\psi_\mu(t)$ function approaches the l_0 norm [6].

The IR scheme. Since ψ is non-convex the problem (2) can be solved by using an IR ℓ_1 scheme [3], where it is substituted by a sequence of convex minimization problems whose solutions $u^{(\ell)}$, $\ell = 1, 2, \ldots$ are proved to converge. The non-convex functional F is locally approximated by its convex majorant F_ℓ obtained by linearizing F at the gradient of an approximate solution $u^{(\ell)}$:

$$F_\ell(Du) = F(Du^{(\ell)}) + f_\ell(Du) - f_\ell(Du^{(\ell)})$$

where $Du^{(\ell)} = (u_x^{(\ell)}, u_y^{(\ell)})$ is the gradient of the image $u^{(\ell)}$ and

$$f_\ell(Du) = \sum_{i,j} \left(\psi'_\mu(|u_x^{(\ell)}|_{i,j})|u_x|_{i,j} + \psi'_\mu(|u_y^{(\ell)}|_{i,j})|u_y|_{i,j} \right).$$

Substituting F with F_ℓ in (2) and choosing $u^0 = \Phi^T y$ we obtain a sequence of approximate solutions:

$$u^{(\ell+1)} = \arg\min_u \mathcal{P}(\lambda, f_\ell(Du), u) \quad \ell = 0, 1, \dots \tag{3}$$

where

$$\mathcal{P}(\lambda, f_\ell(Du), u) = \left\{ f_\ell(Du) + \frac{1}{2\lambda}||\Phi u - z||_2^2 \right\}. \tag{4}$$

The convergence of the sequence $u^{(\ell)}$ defined by (3) to a local minimum of problem (2) is proved in the following theorem.

Theorem 2.1. *The sequence $u^{(\ell)}$ defined by (3) converges to a local minimum of problem (2).*

Proof. We observe that the objective function of (2) satisfies the Kurdika-Lojasiewicz property [7]. In fact

$$\sum_{i,j} (\psi_\mu(|u_x|_{i,j}) + \psi_\mu(|u_y|_{i,j})) + \frac{1}{2\lambda}||\Phi u - z||_2^2$$

is an analytic function, because the function ψ is evaluated in nonnegative arguments, therefore we can restrict the domain of ψ to $\mathbb{R}^+ \bigcup \{0\}$. Finally the thesis immediately follows from [8], where the authors prove the convergence of the IR algorithm when the objective function satisfies the Kurdika-Lojasiewicz property. □

The FNCR Algorithm. In order to avoid sub optimal solutions, due to the presence of many local minima, the parameters μ and λ are decreased on the basis of a continuation strategy. An outer loop, which depends on the index k, is used to decrease the parameter μ from its initial value $\mu \simeq 1$ as follows $\mu = \max(0.8\mu, 10^{-4})$, implementing in this way a graduated non-convexity procedure. Concerning the parameter λ, starting with an initial estimate $\lambda_0 = r_\lambda \|u^{(0)}\|_1$, where $0 < r_\lambda < 1$ is an assigned constant, we propose an iterative update by observing that if $\lambda_k < \lambda_{k-1}$, and $u^{(k)} = \arg\min_u \mathcal{P}(\lambda_k, f_k(Du), u)$, then it easily follows that

$$\mathcal{P}(\lambda_k, f_k(Du^{(k)}), u^{(k)}) < \mathcal{P}(\lambda_{k-1}, f_{k-1}(Du^{(k-1)}), u^{(k-1)})$$

namely, [9]

$$\lambda_{k+1} = \lambda_k \frac{\mathcal{P}(\lambda_k, f_k(Du^{(k)}), u^{(k)})}{\mathcal{P}(\lambda_{k-1}, f_{k-1}(Du^{(k-1)}), u^{(k-1)})}. \tag{5}$$

Moreover a warm starting technique is applied to the iterative solution of (3). These strategies allow us to obtain good quality solutions, avoiding to get stuck in unwanted local minima.

We include all the above observations in the Fast Non Convex Reweighting Algorithm (FNCR) and we report it in **Algorithm 2.1**. The solution of problem (3) is obtained by the iterative Forward-Backward (6)-(7) splitting approach [10] combined with a FISTA (8) acceleration strategy [11]. The forward step (6) depends on the parameter $\beta \in (0, 2/\lambda_{max}(\Phi^t\Phi)]$, where $\lambda_{max}(\Phi^t\Phi)$ is the maximum eigenvalue of the matrix $\Phi^t\Phi$. In MRI CS we have $\lambda_{max} = 1$ hence $\beta \in (0, 2]$. In the present work we set $\beta = 1$ since this value satisfies the necessary condition for the convergence. The Backward step (7) consists of a weighted Total Variation Minimization and for its solution we us the Weighted Split Bregman method [4], where the fidelity term is modified by introducing suitable weights depending from the image edges. The forward-backward iterations are stopped as soon as the following `stopping condition` is true:

$$\Delta_n - \Delta_{n-1} < \gamma\lambda_k, \quad \text{where} \quad \Delta_n = \sum_{i,j=1}^{N} \left((w_{i,j}^x)|\hat{u}_x^{(n)}|_{i,j} + (w_{i,j}^y)|\hat{u}_y^{(n)}|_{i,j} \right). \tag{9}$$

Algorithm 2.1 (INPUT: y, Φ, λ_0, $r_\lambda > 0$, $\gamma > 0$, $\mu > 0$; OUTPUT: λ_{k-1},$u^{(k-1)}$).

FNCR Algorithm

$u^{(0)} = \phi^T z$, $(w^x_{i,j}) = 1$, $(w^y_{i,j}) = 1$, $\lambda_0 > 0$

$\hat{u}^{(0)} = u^{(0)}$, $k = 0$

`repeat` *(IR iteration)*

$t_0 = 1$, $n = 0$

`repeat`

(Forward Step)

$$v = \hat{u}^{(n)} + \beta\Phi^T(z - \Phi\hat{u}^{(n)}) \tag{6}$$

(Backward Step)

$$\tilde{u}^{(n+1)} = \arg \min_{u \in R^{N\times N}} \left\{ \sum_{i,j=1}^{N} \Big((w^x_{i,j})|u_x|_{i,j} + (w^y_{i,j})|u_y|_{i,j} \Big) + \frac{1}{2\beta\lambda_k} ||u - v||_2^2 \right\} \tag{7}$$

(Fista Acceleration)

$t_{n+1} = \frac{1+\sqrt{1+4t_n^2}}{2}$, $\alpha_n = \frac{t_n - 1}{t_{n+1}}$

$$\hat{u}^{(n+1)} = \tilde{u}^{(n+1)} + \alpha_n(\tilde{u}^{(n+1)} - \tilde{u}^{(n)}) \tag{8}$$

$n = n + 1$

`until stopping condition` *as in* (9)

Update λ_{k+1} *as in* (5)

$\mu = \mu \cdot 0.8$ *(μ updating)*

$u^{(k+1)} = \hat{u}^{(n)}$

$\hat{u}^{(0)} = u^{(k+1)}$ *(Warm starting)*

$(w^x_{i,j}) = \psi'_\mu(|u_x^{(k+1)}|_{i,j})$, $(w^y_{i,j}) = \psi'_\mu(|u_y^{(k+1)}|_{i,j})$, *(Weights updating)*

$k = k + 1$

`until convergence`

The parameter γ is a positive constant and its value is important to adapt the precision of the method to the difficulty of the reconstruction problem (small number of samples). In case of strong under-sampling, as reported in the numerical experiments of section 3, we use $\gamma \in [10^{-4}, 10^{-1}]$.

3. Numerical Experiments

In this section we report a few preliminary results aimed to prove the effectiveness of the FNCR algorithm in terms of reconstruction quality and computational efficiency. All the reported experiments are performed on an Intel i7-3770 CPU with 16 GB of RAM, using Matlab R2012b. We are interested in testing FNCR in very critical situations, therefore we choose a difficult test image such as the `forbild` phantom [12] represented in figure 1(a). The synthetic undersampled data z are obtained as $z = \Phi u$ where u is the phantom image and Φ is the Fourier matrix undersampled according to K-space (K) or Radial (R) sampling strategies. In the case of K-space sampling we defined the test problems K16 and K32, obtained by acquiring 16 and 32 vertical lines of the image Fourier spectrum, with a Gaussian random phase encoding. The sampling rates are 12.5% and 25%(figure 1(c)) respectively. The Radial under-sampling is used in the test problems: R9, R10, R12 where 9, 10, 12 radial lines are acquired in the Fourier space with sampling rates 3.7%, 4.1%(figure 1(d)) and 4.8% respectively. In our experiments we focus on the best accuracy achieved by the FNCR algorithm. To this purpose we stop the outer iterations of algorithm 2.1 as soon as $psnr(U)$ reaches its maximum or when $psnr(U) \geq 100$, where U is the image reconstructed by the FNCR algorithm, u is the `forbild` phantom and

$$psnr(U) = 20 \log_{10} \left(\frac{\max_{i,j} u_{i,j}}{rmse(U)} \right), \quad rmse(U) = \frac{1}{N} \sqrt{\sum_{i,j=1}^{N} (u_{i,j} - U_{i,j})^2}.$$

Since in our tests we scale the values of the phantom image u in the interval $[0, 1]$, the case $psnr(U) = 100$ is obtained when $rmse(U) = 10^{-5}$, indicating a reconstruction of extremely good quality. In Table 1 we report the $psnr$ value of the initial image $u^{(0)}$ (column 3) and in columns 4 and 5, the $psnr$ values and the execution times, obtained by algorithm 2.1, with the following input parameters: $\lambda_0 = r_\lambda \|u^{(0)}\|_1$, r_λ in the range $[10^{-3}, 5 \cdot 10^{-3}]$, $\beta = 1$, γ in the range $[5 \cdot 10^{-3}, 10^{-1}]$. The $psnr$ values confirm the excellent accuracy reached by K32, R12 and R10 tests. In the cases K16 and R9, where the under-sampling is very severe, lower accuracy is obtained in both cases with $psnr \simeq 30$. Observing the reconstructed images in figures 2(b) and 3(b) we see that the critical under-sampling mainly affects the reconstruction of the small details (figures 2(c) and 3(c)), preserving good quality in the reconstruction of the coarsest features. To get a feel of the quality of our reconstructions compared to the current literature, we report the results of the Chartrand's Method [CM] in [13], obtained by replacing the anisotropic p-shrinkage (with $p = 0.5$) with the soft-thresholding, in the function `mrics.m` available at `http://www.ece.rice.edu/~tag7/Tom_Goldstein/Split_Bregman_files/mrics.zip`, and setting the parameters in order to obtain the most accurate results. Comparing the columns 4 and 6 in Table 1 we observe that the optimal results are never reached by CM obtaining the best accuracy in test R12 with $psnr \simeq 88$. Furthermore we see that, in the most difficult cases (K16 and R9), both algorithms have quite the same accuracy. Finally comparing the computation times, reported in Table 1 (columns 5 and 7) we see that FNCR is always faster than CM in all the test cases.

			FNCR		CM	
Test	%	$psnr(u^{(0)})$	$psnr(U)$	$sec.$	$psnr(U)$	$sec.$
K32	25%	17.38	100.05	38.5	32.75	118.0
K16	12.5%	13.50	30.70	6.4	30.65	117.0
R12	4.8%	16.36	100.12	8.3	87.80	130.0
R10	4.1%	16.32	100.1	31.9	32.02	136.4
R9	3.7%	15.69	28.41	31.0	26.87	90.5

Table 1: FCNR and CM results.

4. Conclusions

The results obtained in these preliminary tests encourage us to continue a deeper analysis of the algorithm. More synthetic and real data will be analyzed and the sensitivity of the FNCR algorithm to parameters setting and data noise will be investigated. Finally a comprehensive comparison with the methods proposed in the recent literature will be performed.

(a) (b) (c) (d)

Figure 1: (a) Forbild Phantom (256×256); (b) image detail; (c) K16 test; (d) R9 test

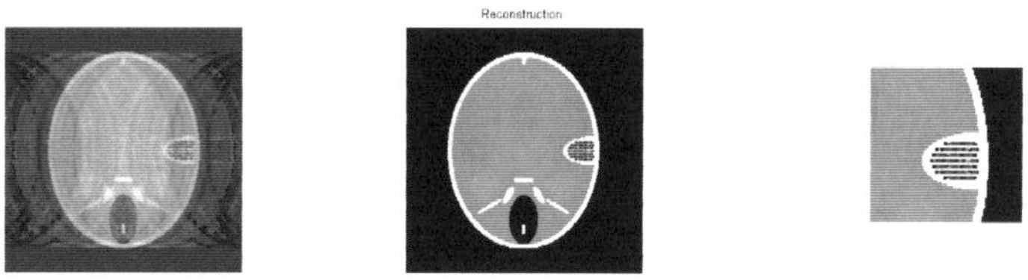

(a) (b) (c)

Figure 2: K16 test. (a) u^0. (b) FNCR reconstruction. (c) Zoom

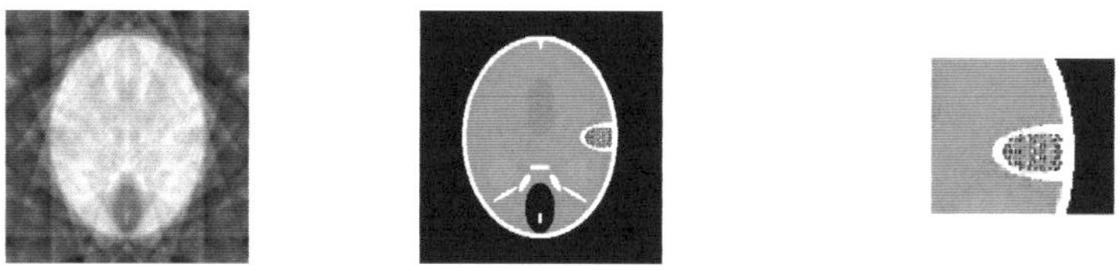

(a) (b) (c)

Figure 3: R9 test.(a) u^0. (b) FNCR reconstruction. (c) Zoom

References

[1] Chartrand R 2007 *IEEE Sig. proc. Lett.* **14** 707–710
[2] Trzasko J and Manduca A 2007 *IEEE Sig. proc. Lett.* **14** 707–710
[3] Candes E and Wakin M 2008 *Journal of Fourier Analysis and Applications* **14** 877–905
[4] Zhang X, Burger M, Bresson X and Osher S 2010 *SIAM J. IMAGING SCIENCES* **3** 253–276
[5] Mourad N and Reilly J 2010 *IEEE Transactions on Signal Processing* **58** 3485–3496
[6] Montefusco L B, Lazzaro D and Papi S 2013 *Signal Processing* **93** 2636 – 2647 ISSN 0165-1684
[7] Attouch H, Bolte J and Svaiter B F 2013 *Math. Program Ser. A* **137** 91–129
[8] Ochs P, Dosovitskiy A, Brox T and Pock T 2015 *SIAM Journal on Imaging Sciences* **8** 331–372
[9] Montefusco L B and Lazzaro D 2012 *IEEE Transactions on Imagel Processing* **21** 1676–1686
[10] Combettes P L and Wajs V R 2005 *Multiscale Modeling & Simulation* **4** 1168–1200
[11] Beck A and Tebulle M 2009 *IEEE Tran. on Imag. Proc.* **18** 2419–2434
[12] Yu Z, Noo F, Dennerlein F, Wunderlich A, Lauritsch G and Hornegger J 2012 *Phys. Med. Biol.* **57**
[13] Chartrand R 2009 Fast algorithms for nonconvex compressive sensing: Mri reconstruction from very few data *IEEE International Symposium on Biomedical Imaging (ISBI)*

Journal of Physics: Conference Series **756** (2016) 012005 doi:10.1088/1742-6596/756/1/012005

Multispectral Electrical Impedance Tomography using Optimization over Manifolds

A Fouchard[1,2], S. Bonnet[1], O. David[2]
[1]Université Grenoble Alpes, CEA, Leti, MINATEC Campus, 17 rue des Martyrs, F38054 Grenoble, Cedex 9, France
[2]Université Grenoble Alpes, Grenoble Institute of Neuroscience, Bâtiment E.J. Safra, Chemin Fortuné Ferrini, F38700 La Tronche, France

E-mail: alexandre.fouchard@centraliens-lyon.net

Abstract. Electrical impedance tomography under spectral constraints uses a material basis decomposition to combine the different information embedded in the tissue spectra. This approach offers an alternative to static imaging while benefiting from systemic error cancellation using difference data. It suits well cases where no prior solution is known and the contrast lies entirely between frequencies, e.g. to diagnose acute stroke or cancer. In this work, a computational framework is presented to deal with the extra frequency dimensions and the constraints during reconstruction. A fraction volume approach is demonstrated with explicit Euclidean gradient, usage of a finite volume element solver and minimization over the oblique manifold. It is applied to synthetic data. Parameter estimations are compared between a mono-frequency inversion and the proposed multispectral implementation. Results suggest that the proposed workflow enables to reduce the computational workload of multispectral inversion while ensuring valid proportions of materials within each control volume.

1. Introduction

Medical electrical impedance tomography (EIT) is a soft-field, non-invasive imaging technique. The setting of EIT consists in reconstructing internal electrical property distributions (conductivity σ and permittivity ε) from current and voltage boundary measurements at specific frequencies ω. The ill-posed EIT inverse problem limits the applications of static imaging, i.e. the deduction of quantitative conductivity maps. Besides, most EIT clinical applications currently use time difference imaging (tdEIT), e.g. in continuous pulmonary and brain function monitoring. It assumes variations of the conductivity distribution along with time, and estimates the parameter changes between two states [1].

The spectral signature of materials may also be used as a source of contrast in frequency difference imaging (fdEIT). This scheme could allow for the imaging without knowledge of a prior condition. It opens application areas in acute stroke, brain injury and cancer screening. However, while tdEIT enables systematic artefact reduction, fdEIT still suffers from strong noise amplification and cannot cope with modelling errors. This significantly reduces the fdEIT range of application [2].

To include frequency information in the image reconstruction process, spectrally constrained EIT (scEIT) has recently been introduced [3]. scEIT then stands as an alternative to static EIT while featuring the advantages of difference imaging techniques. The aim is to reduce the ill-conditioning of the inverse problem. scEIT makes use of a material basis decomposition: the conductivity of a control volume is a linear combination of known tissue conductivities. The reconstruction then focuses on the proportion values of individual tissues within the imaging domain.

Since material proportions are frequency independent, multi-frequency data can be used directly. This contributes to reducing the degrees of freedom of the inversion problem, as long as the number of frequencies exceeds the number of tissues. Further, the knowledge of the tissue spectra enables using difference data in the objective function. Modelling and frequency independent instrumentation errors are mitigated in a nonlinear reconstruction scheme.

The drawbacks of the present approach is the computational workload. The iterative nonlinear inversion workflow involves solving the forward problem, calculating the multi-frequency Jacobian matrix and computing the update step several times. Constraints over proportions have also to be taken into account: they are positive and sum to one for each control volume.

In order to optimize the scEIT imaging method, we propose here a numerical framework based on the joint use of a finite volume element (FVE) forward solver and an optimization on a manifold for the quantitative parameter estimation. FVE brings a model reduction capability compared to finite element methods (FEM) without loss of information [4]. Further, the minimization of the objective function is performed over the oblique manifold to efficiently handle the constraints over proportions [5]. This approach standardizes the integration of spectral constraints. It offers a way to reduce the computational complexity of multi-frequency imaging.

Once the notations of the proportion mixing and the cost functions are fixed, the explicit Euclidean gradient is derived. The methodology of the minimization of the cost function over the oblique manifold is then detailed. Reconstructions over synthetic data are compared between mono-frequency static imaging and the proposed multispectral workflow.

2. Methods

2.1. Material basis: proportion mixing

The parameter estimation under special constraints considers the medium as a mixing model: (i) it is composed of a known number of tissues $t = 1 \dots T$; (ii) the conductivity spectrum s of each tissue is assumed to be known at all frequencies $f = 0 \dots N_f$ of measurements: $s_{ft} = \sigma(\omega_f, t)$; (iii) the conductivity of the kth control volume $k = 1 \dots N$ is fixed by a linear combination of the proportions p_{kt} of the different components, equation (1).

$$\sigma_k(\omega_f) = \sum_t p_{kt} s_{ft} \quad s.t. \quad \begin{array}{c} 0 \leq p_{kt} \leq 1 \\ \sum_t p_{kt} = 1 \end{array} \tag{1}$$

Proportions of a tissue t are gathered under the vector $\mathbf{p}_t \in \mathbb{R}^{N\times 1}$ and considering all tissues in the matrix $\mathbf{P} = [\mathbf{p}_1 \dots \mathbf{p}_T] \in \mathbb{R}^{N\times T}$. In the same way, conductivities are gathered for a frequency f in the vector $\boldsymbol{\sigma}_f = \sigma(\omega_f) \in \mathbb{R}^{N\times 1}$ and in the matrix $\boldsymbol{\Sigma} = \left[\boldsymbol{\sigma}_0 \dots \boldsymbol{\sigma}_{N_f}\right] \in \mathbb{R}^{N\times(N_f+1)}$ for multi-frequency considerations. The mixing model of equation (1) is then expressed as $\boldsymbol{\Sigma} = \mathbf{P}\mathbf{S}^T$. The conductivity at a frequency f is deduced by $\boldsymbol{\sigma}_f = \mathbf{P}\mathbf{s}_{f:}$, with $\mathbf{S} = \left[\mathbf{s}_{0:}^T \dots \mathbf{s}_{N_f:}^T\right]^T \in \mathbb{R}^{(N_f+1)\times T}$. Computing derivatives then makes use of the chain rule and includes the standard Jacobian matrix $\mathbf{J}$. For instance, the derivative of the predicted data, i.e. the computed measurements $\tilde{\mathbf{u}}$, relative to the proportions values: $\frac{\partial \mathbf{u}(\sigma_f)}{\partial \mathbf{p}_t} = \frac{\partial \sigma_f}{\partial \mathbf{p}_t}\frac{\partial \mathbf{u}(\sigma_f)}{\partial \sigma_f} = s_{ft}\mathbf{J}^T(\boldsymbol{\sigma}_f)$. In the following, $\mathbf{I}_n$, $\mathbf{0}_T$ and $\mathbf{1}_n$ denote respectively the identity matrix, the null vector, and the vector of ones of the specified dimension n.

2.2. Cost function

In scEIT, the cost function $\Phi(\mathbf{P})$ is expressed as the weighted sum of a data fit term $\Phi^{(df)}(\mathbf{P})$ and a spatial regularisation term $\Phi^{(r)}(\mathbf{P})$: $\Phi(\mathbf{P}) = \Phi^{(df)}(\mathbf{P}) + \alpha\Phi^{(r)}(\mathbf{P})$. α is a regularization parameter which balances the contribution of each term.

The data fit term $\Phi^{(df)}(\mathbf{P}) = \frac{1}{2}\sum_{f=1}^{N_f}\sum_{m=1}^{M} z_{mf}^2(\mathbf{P})$ is explained in equation (2). $\mathbf{u}(\omega_f)$ denotes the measured voltages at frequency ω_f over all source-detector configurations, indexed by $m = 1 \dots M$, and η a constant independent of proportions.

$$z_{mf} = \frac{\tilde{u}_m(\mathbf{P}\mathbf{s}_{f:}) - \tilde{u}_m(\mathbf{P}\mathbf{s}_{0:})}{\tilde{u}_m(\mathbf{P}\mathbf{s}_{0:})} - \frac{u_m(\omega_f) - u_m(\omega_0)}{u_m(\omega_0)} = \frac{\tilde{u}_m(\mathbf{P}\mathbf{s}_{f:})}{\tilde{u}_m(\mathbf{P}\mathbf{s}_{0:})} + \eta \tag{1}$$

The data fit term takes into account (i) the multi-frequency data, to increase the number of constraints and (ii) the normalized difference data, to reduce modelling and instrumentation errors. ω_0 is chosen as the reference frequency.

2.3. Explicit Euclidean gradient

The Euclidean gradient of the cost function is now derived explicitly.

Following the chain rule, $\frac{\partial \Phi^{(df)}}{\partial \mathbf{P}} = \sum_f \frac{\partial z_{mf}}{\partial \mathbf{P}} \frac{\partial \Phi^{(df)}}{\partial z_{mf}} = \sum_f \frac{\partial z_{mf}}{\partial \mathbf{P}} z_{mf}$. As a quotient derivative, the equation (3) holds, for each control volume k and tissue t.

$$\frac{\partial z_{mf}}{\partial p_{kt}} = \frac{[\mathbf{J}(\mathbf{P}\mathbf{s}_{f:})]_{mk} s_{ft} \tilde{u}_m(\mathbf{P}\mathbf{s}_{0:}) - \tilde{u}_m(\mathbf{P}\mathbf{s}_{f:})[\mathbf{J}(\mathbf{P}\mathbf{s}_{0:})]_{mk} s_{0t}}{u_m^2(\mathbf{P}\mathbf{s}_{0:})} \tag{3}$$

This translates into equation (4).

$$\left[\nabla\Phi^{(df)}(\mathbf{P})\right]_{tk} = \sum_f \sum_m \frac{[\mathbf{J}(\mathbf{P}\mathbf{s}_{f:})]_{mk} s_{ft} \tilde{u}_m(\mathbf{P}\mathbf{s}_{0:}) - \tilde{u}_m(\mathbf{P}\mathbf{s}_{f:})[\mathbf{J}(\mathbf{P}\mathbf{s}_{0:})]_{mk} s_{0t}}{u_m^2(\mathbf{P}\mathbf{s}_{0:})} z_{mf} \tag{4}$$

To simplify the understanding, the following notations are used: $F = N_f - 1$, $\mathbf{z} = \mathrm{vec}(\mathbf{Z}) \in \mathbb{R}^{MF\times 1}$ $\mathbf{s}^{(0)} \in \mathbb{R}^{1\times T}$, $\mathbf{S}^{(1f)} \in \mathbb{R}^{F\times T}$, i.e. $\mathbf{S}^{(1f)}$ contains the piled spectra for the frequencies indexed 1 to F participating to the reconstructions. $\mathbf{a} \in \mathbb{R}^{M\times 1}$, $\mathbf{b} \in \mathbb{R}^{MF\times 1}$, $\mathbf{J}^{(0)}$ and $\mathbf{J}^{(1f)}$ are defined by equation (5).

$$a_m = \frac{1}{\tilde{u}_m(\mathbf{P}\mathbf{s}_{0:})} \qquad \mathbf{b} = \begin{bmatrix} \mathbf{b}^{(1)} \\ \vdots \\ \mathbf{b}^{(F)} \end{bmatrix} \qquad b_m^{(f)} = \frac{\tilde{u}_m(\mathbf{P}\mathbf{s}_{f:})}{u_m^2(\mathbf{P}\mathbf{s}_{0:})} \tag{5}$$

$$\mathbf{J}^{(1f)} = [\mathbf{J}^T(\mathbf{P}\mathbf{s}_{1:}) \;\dots\; \mathbf{J}^T(\mathbf{P}\mathbf{s}_{F:})] \in \mathbb{R}^{N\times MF} \qquad \mathbf{J}^{(0)} = [\mathbf{J}^T(\mathbf{P}\mathbf{s}_{0:}) \;\dots\; \mathbf{J}^T(\mathbf{P}\mathbf{s}_{0:})] \in \mathbb{R}^{N\times MF}$$

Using these notations, the Euclidean gradient can be written in a explicit form as in equation (6), which is suitable for implementation. $\otimes$ denotes the Kronecker product.

$$\nabla\Phi^{(df)}(\mathbf{P}) = \mathbf{J}^{(1f)}\mathrm{diag}(\mathbf{z})\left[\mathbf{S}^{(1f)} \otimes \mathbf{a}\right] - \mathbf{J}^{(0)}\mathrm{diag}(\mathbf{z})\left[\mathbf{s}^{(0)} \otimes \mathbf{b}\right] \tag{6}$$

2.4. Methodology of inversion under constraints

The cost function associated with the estimation of the proportions of the different components is nonlinear. An optimization over a manifold is used to estimate parameters while imposing the constraints on proportions.

2.4.1. Optimization on the oblique manifold The auxiliary variable $\boldsymbol{X} = \sqrt{\boldsymbol{P}^T} \in \mathbb{R}^{N\times T}$, i.e. $p_{kt} = x_{tk}^2$, is introduced. It translates the constraints on proportions into $\sum_t x_{tk}^2 = 1$. Following this change of variable, it amounts to an optimization under equality constraints, which is seamlessly handled by using the oblique manifold $\mathcal{M} = \mathcal{OB}(N_t, N_e) = \{\boldsymbol{X} \in \mathbb{R}^{N\times T} : (\boldsymbol{X}^T\boldsymbol{X})_{ii} = 1, i = 1{:}N\}$ (matrices with unitary column), also ensuring positivity of proportions.

The proportions can be directly estimated within the manifold $\mathcal{M}$. The Matlab® library Manopt, dedicated to manifold optimization, is used [6]. The Euclidean gradient of the objective function is projected onto $\mathcal{M}$, in order to determine the search direction and to evolve within $\mathcal{OB}$ using the concept of retractions.

Journal of Physics: Conference Series **756** (2016) 012005 doi:10.1088/1742-6596/756/1/012005

Dedicated functions are implemented in Matlab® to provide the objective function and its Euclidean gradient, given an estimation of proportions. The chain rule enables deducing the gradient of the objective function relatively to the auxiliary variable **X**: it translates into the Hadamard product of the Euclidean gradient by a factor $2\mathbf{X}^T$. The conjugate-gradient algorithm is then used to estimate the proportions for its efficiency, associated with an optimal line search at each iteration [6].

Regarding the implementation choices, a special care has to be taken for the algorithm initialization. The matrix proportions could be initialized with the first column equal to $\mathbf{p}_1 = \mathbf{1}_N$ (the background of the image) and the other set to $\mathbf{0}_N$. However, using the Manopt library and the change of variable would lead to a null gradient over all objective variables except $\mathbf{p}_1$: the algorithm would remain stuck in the vicinity of the initial point. We propose here to slightly shift from this point to allow for a correct evolution of the algorithm. The initial point is thus set to $\mathbf{p}_1 = 0.9999 * \mathbf{1}_N$ (background) and $\mathbf{p}_t = \frac{0.0001}{T-1} * \mathbf{1}_N \forall t \geq 2$ (other tissues). The associated medium then includes mostly the background and a small proportion of the mixing of all other identified components.

2.4.2. Markov Random Field spatial regularization As in [3], the ill-posed inverse problem is handled by a Markov random field (MRF), equation (7). $l(k)$ designates the neighbors of the control volume k. The objective is to constrain neighboring control volumes to have close proportion values. Such a regularization term provides robustness to noise while allowing a low computational time.

$$\Phi^{(r)}(\mathbf{P}) = \frac{1}{2}\sum_t \sum_k \sum_{l(k)} \left(p_{kt} - p_{l(k)t}\right)^2 \tag{7}$$

This term is efficiently computed using the Laplacian matrix **L** of the graph: $\Phi^{(r)}(\mathbf{P}) = (\mathrm{vec}(\mathbf{P})^T)(\mathbf{I}_T \otimes \mathbf{L})\mathrm{vec}(\mathbf{P})$. Its gradient is given by: $\nabla\Phi^{(r)}(\mathbf{P}) = 2\mathbf{LP}$. The chain rule is then applied to deduce the derivative relatively to the auxiliary variable **X**. The regularization parameter is deduced by the L-method [8]. The latter is applied on the residuals of the cost function data fit term according to the regularization parameter. The graph features two asymptotic lines. The abscissa of their intersection point is chosen as the optimal hyper-parameter.

2.5. FVE forward solver

The proportions are estimated using a FVE method [4,7]. The underlying idea is to shift the focus of the control volume from an elemental resolution to a nodal resolution. Compared to the initial FEM implementation in [3], this allows to reduce the number of control volumes N, e.g. in the study case by a factor 2, i.e. the ratio of the number of elements versus the number of nodes. This is of utmost importance, since the multi-frequency Jacobian computed at each iteration scales with $N \times MF$.

3. Results

A medium with two components – background and conductive inclusion – was considered, and their spectra were assumed known (Figure 1). Synthetic multispectral data was predicted by FEM on a 14 equally-spaced electrode 2D model. It used 15 frequencies, and a 2D forward mesh including 6,450 nodes. Reconstructions were performed using the same numerical phantom, and a different mesh, including 3,527 nodes corresponding to 6,713 elements.

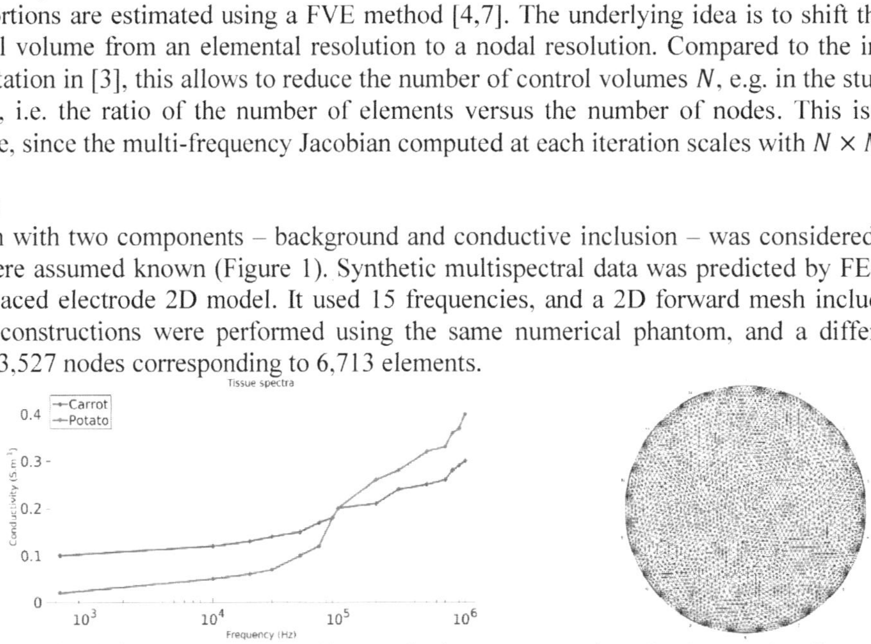

Figure 1. Conductivity spectra assumed known in the reconstruction - background: saline mixed with hashed carrots in blue, inclusion: potato in blue (left); inversion mesh with 14 electrodes (right).

Three situations were compared: (i) the parameter estimation using a mono-frequency nonlinear algorithm (Levenberg-Marquardt), (ii) the inversion through the proposed multispectral nonlinear scheme without spatial regularization, (iii) the deduction of proportions with the regularized multispectral framework together with the choice of an optimal regularization parameter. Conductivity maps and profiles were displayed for the results (the black circle indicated the expected position; the theoretical profiles were in red and the estimated ones in blue).

3.1. Mono-frequency reconstructions

Mono-frequency reconstructions led to estimated conductivity profiles that presented oscillations, due to the ill-posed behaviour of the inverse problem (Figure 2). The contrasts were relatively well recovered by the nonlinear inversion algorithm, relatively to the expected results.

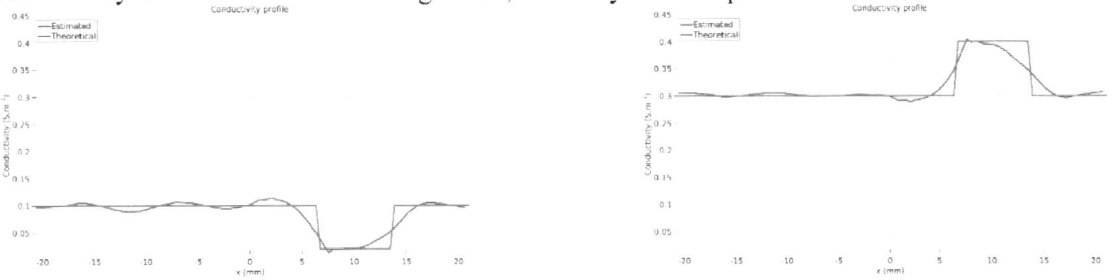

Figure 2. Conductivity profiles from mono-frequency nonlinear reconstructions at 640 Hz and 1 MHz.

3.2. Multispectral inversion using spectral constraints

The proportions of the different components were then estimated using the proposed scheme under spectral constraints. The multi-frequency data enabled to deduce proportion profiles that would no longer present oscillations, i.e. the ill-posed behaviour was reduced (Figure 3). Without spatial regularization, strong reconstruction artefacts were observed, namely in the vicinity of the centre of the numerical phantom. The MRF regularization tended to smooth the obtained profile, which approached the expected one.

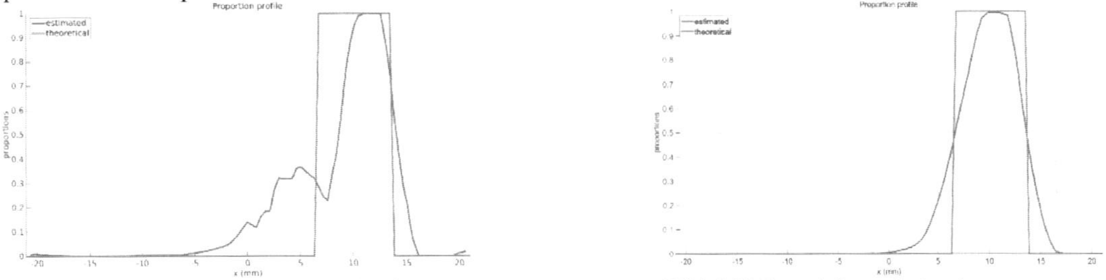

Without spatial regularization With MRF spatial regularization

Figure 3. Proportion profiles deduced by multispectral inversion (scEIT).

The inclusion of the spatial MRF regularization was then considered. Values of the hyper-parameter α were tried between 10^{-5} and 10^{-1} (Figure 4). For low values of α, artefacts were observed in the centre of the imaging domain, but the contrast in proportion was kept. For high values of α, the contrast was sharply inferior to the expected value, and artefacts tended to appear close to the boundary of the domain, with a spatial spread of the recovered proportions.

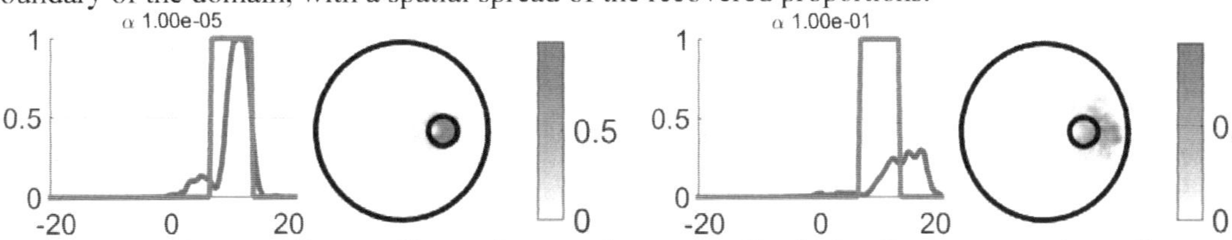

Figure 4. Proportion profiles and proportion maps of inclusion for extreme values of $\boldsymbol{\alpha}$.

Mono-frequency conductivity maps were then compared with spectrally constrained proportions (Figure 5). The former case showed that for some frequencies, the inclusion was not discernible; ringing effects were visible. The latter case showed a properly reconstructed inclusion: the spectral constraints allowed to feature very few artefacts and low ringing effects.

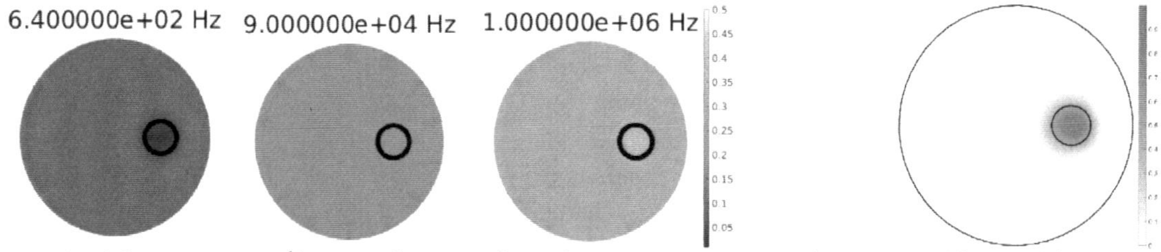

Conductivity maps ($S.m^{-1}$), mono-frequency inversion Proportion map, multispectral inversion

Figure 5. Estimated reconstructions from synthetic data.

4. Discussion

In this paper, we presented a numerical framework to handle the EIT multispectral inversion under spectral constraints. The multi-frequency Euclidean gradient was derived explicitly. Proportions were then estimated using the oblique manifold and a FVE solver (instead of standard FEM) to reduce the computational workload. The optimal regularization parameter was chosen using a bilinear regression.

Methods were tested on synthetic data. As expected, compared with a mono-frequency implementation, the quality of the multispectral reconstruction was enhanced. In this scope, the framework allowed to standardize the inversion workflow with the Manopt library. The minimization under equality constraints allowed ensuring valid proportions of materials at each control volume.

The proposed scEIT framework also enabled to enhance the efficiency of the inversion process, through the use of a FVE forward solver. With the 14 electrode phantom and the 15 frequencies used in this work, the size of the multi-frequency Jacobian that was computed at each iteration (20 iterations on average) was halved, translating into effective computation and assembly time savings. It would have valued in FEM $6{,}713 \times 14^2 * (15-1)$ elements versus $3{,}527 \times 14^2 * (15-1)$ in FVE.

Further studies will focus on applying these numerical developments to *in vitro* and *in vivo* data, with a straightforward extension to 3D and numerous tissues. They will also consider the robustness to the noise in the data and to the variations of the contact impedance at the medium-electrode junction.

References

[1] Holder D S (ed) 2004 *Electrical Impedance Tomography: Methods, History, and Applications* (London: Taylor & Francis)

[2] Seo J K and Woo E J 2013 Electrical impedance tomography *Nonlinear Inverse Problems in Imaging* (John Wiley & Sons, Ltd) 195–249

[3] Malone E, Sato dos Santos G, Holder D S and Arridge S 2014 Multifrequency electrical impedance tomography using spectral constraints *IEEE Trans. Med. Imaging* **33** 340–50

[4] Fouchard A, Bonnet S and David O 2015 Méthodes numériques pour le problème direct et l'analyse de sensibilité en tomographie d'impédance électrique *GRETSI* 2015 (Lyon, France)

[5] Absil P-A, Mahony R and Sepulchre R 2009 *Optimization Algorithms on Matrix Manifolds* (Princeton University Press)

[6] Boumal N, Mishra B, Absil P-A and Sepulchre R 2014 Manopt, a Matlab toolbox for optimization on manifolds *Journal. Mach. Learn. Res.* **15** 1455–9

[7] Fouchard A, Bonnet S, Hervé L and David O 2015 A current-density conservative nodal framework for EIT *International Conference on Biomedical Applications of Electrical Impedance Tomography* (Neûchatel, Suisse)

[8] Salvador S and Chan P 2004 Determining the number of clusters/segments in hierarchical clustering/segmentation algorithms *16th IEEE International Conference on Tools with Artificial Intelligence* 576–84 (Boca Raton, Florida, USA)

A new nonconvex variational approach for sensory neurons receptive field estimation

A. Drogoul[1], G. Aubert[2], B. Cessac[1], P. Kornprobst[1]

[1] Inria, Biovision team, Sophia Antipolis, France

[2] Laboratoire J.A. Dieudonné, UMR 6621 CNRS Université Nice Sophia Antipolis, France

E-mail: audric.drogoul@inria.fr

Abstract. Determining the receptive field of a visual sensory neuron is crucial to characterize the region of the visual field and the stimuli this neuron is sensitive to. We propose a new method to estimate receptive fields by a nonconvex variational approach, thus relaxing the simplifying and unrealistic assumption of convexity made by standard approaches. The method consists of solving a relaxed discrete energy minimization problem using a proximal alternating algorithm. We compare our approach with the classical spike-triggered-average technique on simulated data, considering a typical retinal ganglion cell as ground truth. Results show a high improvement in terms of accuracy and convergence with respect to the duration of the experiment.

1. Introduction

This paper deals with the estimation of receptive fields of individual visual neurons. Knowing the receptive field of a particular neuron is a crucial information to understand which region of the visual field and which stimulus the neuron is sensitive to. Thus, biologists always devote efforts in their experimental protocol to characterize the neurons they are recording.

The goal is to characterize the relation between the stimulus and the neuron response which is a set of action potentials (also called spikes). To do so, it is classical to assume that response follows a linear-nonlinear Poissonian (LNP) model [6, 9, 12] (Fig. 1): given a visual stimulus $s(x,t) : \Omega \subset \mathbb{R}^2 \times [0,T] \to \mathbb{R}$, where Ω is the spatial domain and T is the duration of the experiment, the neuron generates a sequence of $n(T)$ spikes times $\{t_i\}_{1\leq i\leq n(T)}$ such that

$$\{t_i\}_{1\leq i\leq n(T)} \text{ is generated by a Poisson process of rate } r(t) = \mathcal{S}\left((s\times u)(t)\right), \tag{1}$$

where $\mathcal{S}(.)$ is a nonlinear function and $u : \Omega_d = \Omega \times [-d, 0] \longrightarrow \mathbb{R}$ is the so-called receptive field which corresponds to the linear part of the processing, where $d > 0$ is the length of its temporal support.

The operator $\times$ is defined by $(s\times u)(t) = \int_{-\infty}^{0} \langle s(.,t+\tau), u(.,\tau)\rangle_\Omega d\tau$, where $< .,. >_\Omega$ denotes the inner product on Ω. Given hypothesis (1) the problem of estimating the receptive field $u(x,t)$ is an inverse problem: given a stimulus $s(x,t)$ of duration T and the $n(T)$ spikes $\{t_i\}_{1\leq i\leq n(T)}$ from the sensory neuron responding to $s(x,t)$, recover the unknown receptive field $u(x,t)$.

This inverse problem can be formulated using a Bayesian approach. Using the Bayes rule, a classical method is to search for the receptive field $u(x,t)$ as minimum of the anti-log of the *a posteriori* probability (see, e.g., [12, 14, 13]):

$$\inf_u \Big\{ -log\left(\rho(u|\{t_i\}_{1\leq i\leq n(T)})\right) \Big\} = \inf_u \Big\{ -log(\rho(\{t_i\}_{1\leq i\leq n(T)}|u)) - log(\rho(u))\Big\}, \tag{2}$$

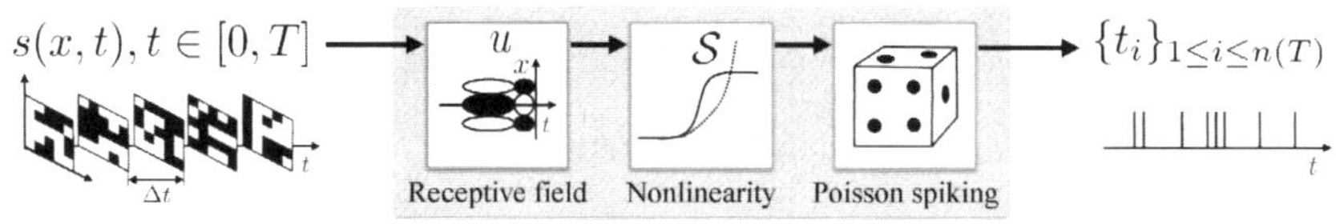

Figure 1. Illustration of the LNP model (1). Two kinds of $\mathcal{S}(.)$ are illustrated: ramp- or sigmoid-like nonlinearities. Here the stimulus is a white noise sequence which is the classical stimulus used for the spike-triggered averaged approach (STA, see [6]).

where $\rho(u)$ denotes the probability density of the random variable u. Then, assuming that $\rho(u)$ is a distribution of the form $\rho(u) \propto e^{-J(u)}$, one can show that (2) can be rewritten as [7]:

$$\inf_u \ \mathcal{E}(u) = \psi(s \times u) + J(u), \tag{3}$$

where $\psi(s \times u) = -log(\rho(\{t_i\}_{1\leq i\leq n(T)}|u))$ is the data fidelity term defined by

$$\psi(z) = \int_0^T \mathcal{S}(z(\tau))d\tau - \sum_{i=1}^{n(T)} log(\mathcal{S}(z(t_i))), \tag{4}$$

where $z(t) = (s \times u)(t)$, and $J(u)$ is the prior term to infer qualitative properties to the solution, according to what is known about the general shape of a receptive field. To study this problem, it is classical to assume that the nonlinearity $\mathcal{S}(.)$ is a ramp function or an exponential. The computation of (3) is then simple since it is a convex problem to solve [12, 14, 9].

On the opposite, in this paper, we consider the case when the nonlinearity $\mathcal{S}(.)$ is a sigmoid, which is more realistic from a physiological point of view; in particular it models the fact that the neuron has a bounded firing rate. However, since $\mathcal{S}(.)$ is nonconvex, the data fidelity term $\psi(s \times u)$ also becomes nonconvex. This is known to be harder to minimize numerically.

Concerning $J(.)$, we choose it to impose two properties on the solution. The first is that u should be localized in space and time since neurons are sensitive to a particular region of the visual field. This can be imposed by a sparsity constraint term. Here we choose to use a convex relaxation of the sparsity and we penalise the L^1-norm of u. The second is that u should be smooth. We propose that u should belong to the space BV_2 that contains piecewise linear functions. This kind of regularity constraint has also been previously used in the context of image restoration [4]. This allows to recover functions with fast smooth variations more precisely than using a simple $\|\nabla u\|^2_{L^2}$-regularity constraint. So the prior term will be

$$J(u) = \lambda \|u\|_{L^1(\Omega_d)} + \mu |u|_{BV_2(\Omega_d)}, \tag{5}$$

where $\lambda, \mu > 0$ are weights associated respectively to the sparsity and to the regularity of u.

So this paper is about the study of the problem (3) with (4)–(5) In Sect. 2, we study the discrete version of the problem (3). We introduce a relaxed problem so that the solution can be computed more easily. We justify it theoretically and introduce an alternated minimizing algorithm converging toward a critical point of the relaxed energy. In Sect. 3, we test the approach on simulated data to provide a quantitative evaluation with comparisons to the classical spike triggered averaged technique (STA, see [6]). In Sect. 4, we describe future work.

2. Study of the discrete problem: Well-posedness and algorithm

2.1. Problem definition

In this section, we study a discrete version of (3). The stimulus is a sequence of N_t images each presented during a period of Δt so that the duration of the experiment is $T = N_t\Delta t$. Each image is of size $N_x \times N_y$ pixels. Receptive field is of size $N_x \times N_y \times D$ where D is a fixed depth. We introduce the real vector spaces $\mathcal{X}$, $\mathcal{Y}$ and $\mathcal{Z}$ so that:

$$u \in \mathcal{X} = \mathbb{R}^{N_x \times N_y \times D}, \quad s \in \mathcal{Y} = \mathbb{R}^{N_x \times N_y \times N_t}, \quad z \in \mathcal{Z} = \mathbb{R}^{N_t}, \tag{6}$$

endowed with the scalar product $\langle .,. \rangle$ and the associated norm $\|.\|_2 = \sqrt{\langle .,. \rangle}$. We denote by $\|.\|_1$ the l_1–norm. We do the following hypothesis on the sigmoid function $\mathcal{S}(.)$.

Hypothesis 1. (i) *$\mathcal{S}$ is semialgebraic [1], bounded (values in $[m_\mathcal{S}, M_\mathcal{S}]$), and non-decreasing;*
(ii) *$\mathcal{S} \in C^0(\mathbb{R})$ and there exists $\rho > 0$ such that $\mathcal{S}(z) + \rho z^2$ is convex;*
(iii) *$z \mapsto -log(\mathcal{S}(z))$ is convex on the set $\{\mathcal{S} > 0\}$.*

Given these notations, if we denote $\xi = (\xi_i)_{1 \le i \le N_t}$ the number of spikes per time step and if we assume that $\Delta t = 1$ without loss of generality, then the data fidelity term (4) can be written as

$$\psi_\xi(z) = \sum_{i=1}^{N_t} \psi_{\xi_i}(z_i), \quad \text{with} \quad \psi_{\xi_i}(z_i) = \mathcal{S}(z_i) - \xi_i log(\mathcal{S}(z_i)), \tag{7}$$

where $\psi_{\xi_i}(z)$ has a limited domain of definition denoted by $\mathcal{Z}^+_{\xi_i}$ equal to $\mathbb{R}$ if $\xi_i = 0$, and $\{\mathcal{S} > 0\}$ if $\xi_i > 0$. One can easily verify that the data fidelity term is nonconvex when $\xi_i \le c$.

The prior term is approximated by

$$J_\varepsilon(u) = \lambda \|u\|_1 + \mu I_\varepsilon(u) \quad \text{with} \quad I_\varepsilon(u) = \sum_{(i,j,k)} \sqrt{\varepsilon^2 + |(Hu)_{i,j,k}|^2}, \tag{8}$$

where $H : \mathcal{X} \to \mathcal{X}^9$ is a matrix equal to the discrete Hessian and $\varepsilon > 0$ is small enough to make the regularization term differentiable.

Remark 2.1. *The introduction of ε could be avoided by introducing a dual variable following for example [3]. However, we did not use this approach because of the computational time since this would imply to solve a primal dual problem and compute a projection on the set $\{H^\star p,\ p \in \mathcal{X}^9,\ |p_{i,j,k}| \le 1\}$ by a fixed point algorithm based on [5].*

Denoting by $\times_d$ the discretization of $\times$ we get the discrete problems associated to (3):

$$\inf_{u \in \mathcal{X}} \mathcal{E}(u) = \psi_\xi(s \times_d u) + i_{\mathcal{Z}^+_\xi}(s \times_d u) + \lambda \|u\|_1 + \mu I_\varepsilon(u), \tag{9}$$

where $i_{\mathcal{Z}^+_\xi}$ denotes the characteristic function associated to $\mathcal{Z}^+_\xi = \prod_{i=1}^{N_t} \mathcal{Z}^+_{\xi_i}$.

In (9), the data fidelity term depends only on $s \times_d u$ and when minimizing w.r.t. u it is difficult to guarantee that $s \times_d u$ remains in the domain of definition of ψ_ξ. To solve this problem, we propose to introduce an auxiliary variable z and the following relaxed formulation:

$$\inf_{z \in \mathcal{Z}, u \in \mathcal{X}} \mathcal{E}_\alpha(z, u) = \psi_\xi(z) + i_{\mathcal{Z}^+_\xi}(z) + \frac{\alpha}{2} \|s \times_d u - z\|_2^2 + \lambda \|u\|_1 + \mu I_\varepsilon(u), \tag{10}$$

where the term in α penalizes the difference between z and $s \times_d u$. Another interest of this relaxed problem is that now the problem in z containing the nonconvexity is separable (it leads to N_t one-dimensional independent problems) and we will show that the energy is separately convex w.r.t. to z and u under some assumptions.

2.2. *Well-posedness*

Assuming that $\mathcal{S}$ verifies Hypothesis 1, we start with the following two propositions which result from standard arguments:

Proposition 2.1 (Existence). *Let $\alpha > 0$ and $\lambda > 0$ (resp. $\lambda > 0$), the energy $\mathcal{E}_\alpha(z, u)$ (10) (resp. $\mathcal{E}(u)$ (9)) is coercive and admits at least a minimizer.*

Proposition 2.2. *If $\alpha > \rho$ then problem $\mathcal{E}_\alpha(z, u)$ (10) is convex w.r.t to z and u separately (but not convex w.r.t. (z, u)).*

The following proposition establishes the link between solutions of the relaxed problem and solutions of the initial one.

Proposition 2.3. *Let α be a positive integer, then the sequence $\{(z_\alpha, u_\alpha)\}_{\alpha\geq 1}$ solution of the minimization problem (10) associated to $\mathcal{E}_\alpha(z, u)$ is bounded and*

(i) *All its cluster points are couples $(s \times \bar{u}, \bar{u})$ such that $\bar{u}$ is a solution of (9).*

(ii) *The infimum converges: $\inf_{(z,u)} \mathcal{E}_\alpha(z, u) = \mathcal{E}_\alpha(z_\alpha, u_\alpha) \underset{\alpha\to+\infty}{\longrightarrow} \inf_u \mathcal{E}(u)$.*

(iii) *If $\mathcal{E}(u)$ admits a unique minimizer $\bar{u}$, then $(z_\alpha, u_\alpha) \underset{\alpha\to+\infty}{\longrightarrow} (s \times \bar{u}, \bar{u})$.*

Proof. The proof of this proposition uses the epi-convergence of $\mathcal{E}_\alpha(z, u)$ toward $\mathcal{E}(u)\ i_{\{z=s\times_d u\}}$ (see [15]). □

2.3. *Algorithm*

To compute a solution of (10), we propose the proximal alternated minimization introduced in [1] in a general context. Given an initial condition $(z^{(0)}, u^{(0)}) \in \mathcal{Z} \times \mathcal{X}$, the algorithm consists of the following two steps:

$$z^{(k+1)} \in \arg\min_z\ \mathcal{E}_\alpha(z, u^{(k)}) + \frac{1}{2\beta^{(k)}}\|z - z^{(k)}\|_2^2, \tag{11a}$$

$$u^{(k+1)} \in \arg\min_u\ \mathcal{E}_\alpha(z^{(k+1)}, u) + \frac{1}{2\gamma^{(k)}}\|u - u^{(k)}\|_2^2, \tag{11b}$$

where $\beta^{(k)}, \gamma^{(k)}$ are sequences of parameters belonging to $[r^-, r^+]$ with $0 < r^- < r^+$ for all $k \geq 0$ (note that this is the only condition on these parameters to obtain convergence). Note that quadratic terms in (11) are necessary to show the convergence and that the definition of (11) leads to a decrease of the energy verifying:

$$\mathcal{E}_\alpha(z^{(k+1)}, u^{(k+1)}) + \frac{1}{2\beta^{(k)}}\|z^{(k+1)} - z^{(k)}\|_2^2 + \frac{1}{2\gamma^{(k)}}\|u^{(k+1)} - u^{(k)}\|_2^2 \leq \mathcal{E}_\alpha(z^{(k)}, u^{(k)}).$$

From a numerical point of view, this algorithm becomes easy to tackle:

- If we choose $\mathcal{S}(.)$ as a piecewise cubic function verifying Hypothesis 1, when $\xi_i = 0$, then problem (11a) can be solved analytically thanks to its separability: it is equivalent to compute the proximal operator [10] of $\mathcal{S}(.)$ up to a multiplicative constant depending on α and $\beta^{(k)}$. When $\alpha + \frac{1}{\beta^{(k)}} > \rho$ the problem is strictly convex so that the proximal operator is uni-valued. Otherwise, in the general case, we use a Newton algorithm.
- Problem (11b) is convex and we solve it by a standard forward-backward proximal algorithm of type Nesterov [2].

Theorem 2.1. (Convergence) *Algorithm (11a)–(11b) generates a sequence $(z^{(k)}, u^{(k)})_{k\in\mathbb{N}}$ which converges to a critical point $(\bar{z}, \bar{u})$ of $\mathcal{E}_\alpha(z, u)$ (i.e., $0 \in \partial\mathcal{E}_\alpha(\bar{z}, \bar{u})$) and the sequence $(z^{(k)}, u^{(k)})$ is $l_1(\mathbb{N})$.*

Proof. Energy $\mathcal{E}_\alpha(z, u)$ verifies the Kurdyka-Lojasiewicz property (see [8] and Definition 7 in [1]) and is coercive from Proposition 2.1. By applying Theorem 9 of [1] we get the result. □

3. Numerical results

In this section we show quantitative results obtained using simulated data. We assume that we look for the receptive field of a neuron which can be approximated by a LNP model and that the nonlinearity is a known sigmoid (a piecewise cubic approximation verifying Hypothesis 1). Here we choose as the ground truth a receptive field shape that corresponds to a class of retinal ganglion cells: the function u is separable in space and time, it is a difference of Gaussian in space and a difference of exponential in time. It is discretized as a spatio-temporal volume of size $20 \times 20 \times 30$ pixels. This is illustrated in Fig. 2 (first row, sample temporal slices).

For the stimulus, we choose a sequence of $N_t = 1000$ uniform random binary images, each presented during a time step $\Delta t = 1$. Images are of size 20×20 pixels, with block size 4×4 pixels (as illustrated in Fig. 1). Spikes are generated by (1), giving around $n(T) = 500$ spikes.

In Fig. 2 we show a comparison between STA [6] and ours. We chose our parameters empirically: $\beta^{(k)} = \gamma^{(k)} = 10\ \forall k$, $\lambda = 10$, $\epsilon = 10^{-8}$, $\mu = 100$) and for the initial condition: $u^{(0)} = 0$, $z^{(0)} = \mathcal{S}^{-1}((m_{\mathcal{S}} + M_{\mathcal{S}})/2)$. Results show that the spatial resolution of STA is constrained by the size of the block of the stimulus, by definition of the STA (PSNR=22.1-dB), while we obtain a much better estimation (PSNR=33.1dB).

Concerning the convergence, different aspects are shown in Fig. 3. In Fig. 3(left), we show

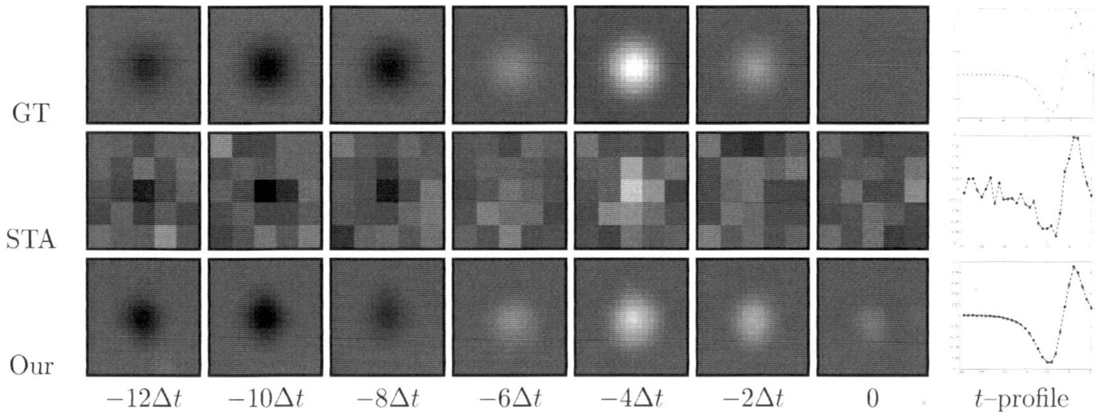

Figure 2. Example of result for simulated data. Each column shows a sample of temporal slices of (first line) the ground truth (GT), i.e., the receptive field to recover, (second line) the result obtained using the classical STA approach, and (third line) the result obtained using our approach. Last column shows temporal profile corresponding to a central position.

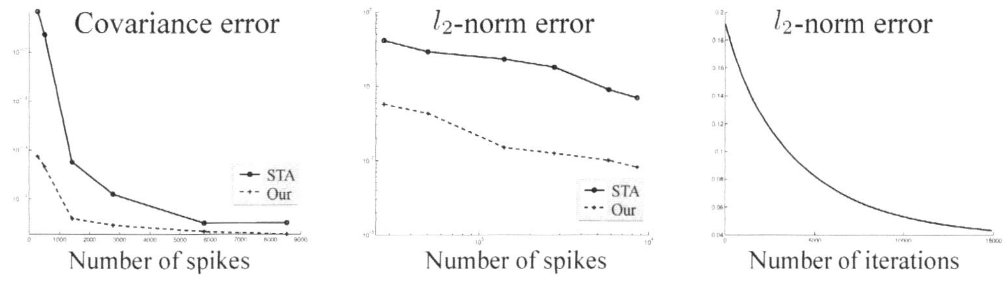

Figure 3. Convergence of the approach: (left) covariance error ($1 - \frac{cov(u,u_0)}{\sigma_u \sigma_{u_0}}$) as a function of the number of spikes in semi-log scale, (middle) l_2-norm error ($\|u - u_0\|_2$) as a function of the number of spikes in log-log scale, and (right) l_2-norm error as a function of the number of iterations for $N_t = 1000$.

that our approach converges faster than STA as the number of spikes increases, which is directly related to the duration of the experiment. In other words, for a fixed duration of experiment, our approach provides a better estimate of the receptive field than STA. In Fig. 3(middle), the main observation is that error decays linearily (in log scale) with a slope of the convergence equal for both methods which can be explained by the central limit Theorem. Finally, Fig. 3(right) illustrates the speed of convergence of our algorithm.

4. Conclusion

Overall, this study presents for the first time an approach to deal with the non-convex case allowing efficient extraction of receptive fields with great accuracy, as exemplified on synthetic data. Another advantage of this approach is that we are no longer constrained to use white noise as input as needed in STA, so that one can investigate what could be an optimal stimulus (e.g., [11]) to further improve the results. Future work will focus on validating our approach on real cell recordings.

Acknowledgements This work was partially supported by the EC IP project FP7-ICT-2011-9 no. 600847 (RENVISION). We also thank Daniela Pamplona (Inria, Biovision team) as well as Evelyne Sernagor and Gerrit Hilgen (Institute of Neuroscience, Newcastle University, UK) for their helpful insights.

References

[1] H. Attouch, J. Bolte, P. Redont, and A. Soubeyran. Proximal alternating minimization and projection methods for nonconvex problems: An approach based on the Kurdyka-Lojasiewicz inequality. *Math. Oper. Res.*, 35(2):438–457, May 2010.

[2] A. Beck and M. Teboulle. Fast Gradient-Based Algorithms for Constrained Total Variation Image Denoising and Deblurring Problems. *IEEE Transactions on Image Processing*, 18(11):2419–2434, November 2009.

[3] J. Bect, L. Blanc-Feraud, G. Aubert, and A. Chambolle. A l1-unified variational framework for image restoration. In *Proceedings of the 8th European Conference on Computer Vision*, page 113, 2004.

[4] M. Bergounioux and L. Piffet. A Second-Order Model for Image Denoising. *Set-Valued and Variational Analysis*, 18(3-4):277–306, December 2010.

[5] A. Chambolle. An Algorithm for Total Variation Minimization and Applications. *J. Math. Imaging Vis.*, 20(1-2):89–97, January 2004.

[6] E.J. Chichilnisky. A simple white noise analysis of neuronal light responses. *Network: Computation in Neural Systems*, 12(2):199–213, 2001.

[7] A. Drogoul, G. Aubert, B. Cessac, and P. Kornprobst. A new nonconvex variational approach for sensory neurons receptive field estimation. Technical Report 8837, Inria, March 2016.

[8] K. Kurdyka. On gradients of functions definable in o-minimal structures. *Annales de l'institut Fourier*, 48(3):769–783, 1998.

[9] J.M. McFarland, Y. Cui, and D. A. Butts. Inferring Nonlinear Neuronal Computation Based on Physiologically Plausible Inputs. *PLoS Computational Biology*, 9(7):e1003143, July 2013.

[10] J.J. Moreau. Proximité et dualité dans un espace hilbertien. *Bulletin de la Société Mathématique de France*, 93:273–299, 1965.

[11] D. Pamplona, G. Hilgen, S. Pirmoradian, M.H. Hennig, B. Cessac, E. Sernagor, and P. Kornprobst. A super-resolution approach for receptive fields estimation of neuronal ensembles. Computational Neuroscience (CNS), July 2015.

[12] L. Paninski. Maximum likelihood estimation of cascade point-process neural encoding models. *Network: Computation in Neural Systems*, 15(4):243–262, November 2004.

[13] M. Park and J. W. Pillow. Receptive Field Inference with Localized Priors. *PLoS Computational Biology*, 7(10):e1002219, October 2011.

[14] J. Pillow. Likelihood-based approaches to modeling the neural code. *Bayesian brain: Probabilistic approaches to neural coding*, pages 53–70, 2007.

[15] R.T. Rockafellar and R.J.B. Wets. *Variational Analysis*, volume 317 of *Grundlehren der mathematischen Wissenschaften*. Springer Berlin Heidelberg, Berlin, Heidelberg, 1998.

Total Variation Denoising and Support Localization of the Gradient

A. Chambolle[1], **V. Duval**[2], **G. Peyré**[2,3] **and C. Poon**[2,3,4]

[1] CNRS and CMAP, Ecole Polytechnique 91128 Palaiseau Cedex, FRANCE
[2] INRIA, MOKAPLAN, Domaine de Voluceau Le Chesnay, FRANCE
[3] CNRS and CEREMADE, Université Paris-Dauphine, Place du Marechal De Lattre De Tassigny, 75775 Paris 16, FRANCE
[4] DAMTP, University of Cambridge, Wilberforce Road, Cambridge CB3 0DZ, UK

E-mail: antonin.chambolle@polytechnique.fr, vincent.duval@inria.fr, {gabriel.peyre, clarice.poon}@ceremade.dauphine.fr

Abstract. This paper describes the geometrical properties of the solutions to the total variation denoising method. A folklore statement is that this method is able to restore sharp edges, but at the same time, might introduce some staircasing (i.e. "fake" edges) in flat areas. Quite surprisingly, put aside numerical evidences, almost no theoretical result are available to backup these claims. The first contribution of this paper is a precise mathematical definition of the "extended support" (associated to the noise-free image) of TV denoising. This is intuitively the region which is unstable and will suffer from the staircasing effect. Our main result shows that the TV denoising method indeed restores a piece-wise constant image outside a small tube surrounding the extended support. Furthermore, the radius of this tube shrinks toward zero as the noise level vanishes and in some cases, an upper bound on the convergence rate is given.

1. Introduction

The total variation (TV) denoising method was introduced by Rudin, Osher and Fatemi in [1]. It is one of the first proposed non-linear image restoration method, and has had an enormous impact on shaping modern imaging sciences. Despite being quite old, this method is still routinely used today, and its popularity probably stems from both its simplicity and its ability to restore "cartoon-looking" images.

Total Variation Denoising The total variation of a function $u \in \mathrm{L}^2(\mathbb{R}^2)$ is defined as

$$J(u) \stackrel{\text{def.}}{=} \int_{\mathbb{R}^2} |Du| \stackrel{\text{def.}}{=} \sup\left\{ \int_{\mathbb{R}^2} u \operatorname{div} z \;;\; z \in \mathscr{C}_c^1(\mathbb{R}^2, \mathbb{R}^2), \|z\|_\infty \leqslant 1 \right\}. \tag{1}$$

Given some noisy input function f, following [1], we are interested in the total variation denoising problem

$$\min_{u \in \mathrm{L}^2(\mathbb{R}^N)} \lambda J(u) + \frac{1}{2}\|u - f\|_{L^2}^2. \tag{$\mathcal{P}_\lambda(f)$}$$

Here, $\lambda > 0$ is the regularization parameter, and it should adapted by the user to the noise level.

In this paper, we present the results of [2], which studies the ability of TV denoising to restore the geometrical structures (in particular the edges) of some (typically unknown) noise-free function f by solving $\mathcal{P}_\lambda(f+w)$. Here w accounts for some additive noise in the image formation process, and is assumed to have a finite L^2 norm $\|w\|_{L^2}$.

Level lines in the low noise regime Although it is not very difficult to see that, as $\lambda \to 0^+$ and $\|w\|_{L^2} \to 0^+$, the solution $u_{\lambda,w}$ to $\mathcal{P}_\lambda(f+w)$ converges towards f in the L^2 topology, our goal is to describe this convergence more precisely: *is it possible to say that the level lines of $u_{\lambda,w}$ converge to those of f? In what sense? Moreover, does the support of $Du_{\lambda,w}$ converge towards the support of Df?*

1.1. *Previous works*

Jump sets stability A landmark result in the characterization the structural properties of TV denoising solutions is [3], which proves that TV regularization does not introduce jumps, i.e. the "jump set" of the solution of $(\mathcal{P}_\lambda(f))$ is included in that of the input f. A review of this result and extensions can be found in [4].

These results are however of little interest when f is replaced by a noisy function $f+w$, since the noise w, which is only assumed to be in L^2, might introduce jumps everywhere. It is actually the presence of this noise which is responsible for the "staircasing" effect, which creates spurious edges in flat area. The present paper aims to fill this theoretical gap by analyzing the impact of the noise on the gradient support of the solution to $\mathcal{P}_\lambda(f+w)$, when both $\|w\|_{L^2}$ and λ are not too large.

Explicit solutions and calibrable sets Of particular importance for the analysis of TV methods are indicator functions of sets, and their behavior under the regularization. Indicator functions which are invariant (up to a rescaling) under TV denoising define the so-called "calibrable" sets. More precisely, by denoting the perimeter of a set by $P(C) \stackrel{\text{def.}}{=} J(\mathbb{1}_C)$ and its Lebesgue measure by $|C|$, we say that $C \subset \mathbb{R}^2$ is a calibrable set if $h_C \mathbb{1}_C \in \partial J(\mathbb{1}_C)$, where $h_C = P(C)/|C|$; and if C is a calibrable set, then the solution of $\mathcal{P}_\lambda(\mathbb{1}_C)$ is $u_{\lambda,0} = (1-h_C\lambda)^+\mathbb{1}_C$. A full characterization of convex calibrable sets is given in [5]. It was also proved in [5] that if C is a convex set, then the solution u_λ to $\mathcal{P}_\lambda(\mathbb{1}_C)$ satisfies

$$\{u_\lambda \geqslant t\} = \emptyset \text{ if } t > 1 - \frac{\lambda}{R^*} \quad \text{and} \quad \{u_\lambda \geqslant t\} = C_{\lambda/(1-t)} \text{ if } 0 \leqslant t \leqslant 1 - \frac{\lambda}{R^*}$$

where $C_r = \bigcup\{B(x,r)\ ;\ B(x,r) \subseteq C\}$ and C_{R^*} is the maximal calibrable set in C. Throughout this paper, $B(x,r)$ denotes the ball of radius r centred at x. In this paper, we shall show that in the presence of L^2 additive noise, calibrable sets play the role of "stable" sets and the corresponding edges are well restored by TV denoising.

Integral error estimates The systematic study of noise stability of regularization schemes relies on the so-called *source condition* [6], which in the simple denoising setting, is the requirement that the subdifferential of J, $\partial J(f)$, should be non-empty. In the case of total variation regularization, the subdifferential of $J : \mathrm{L}^2(\mathbb{R}^2) \to \mathbb{R} \cup \{+\infty\}$ can be written as follows.

$$\partial J(0) = \left\{\operatorname{div} z\ ;\ z \in L^\infty(\mathbb{R}^2;\mathbb{R}^2),\ \operatorname{div} z \in L^2(\mathbb{R}^2),\ \|z\|_{L^\infty} \leqslant 1\right\},$$
$$\partial J(u) = \left\{v \in \partial J(0)\ ;\ \int_{\mathbb{R}^2} uv = J(u)\right\}.$$

For non-smooth regularizations over Banach spaces, stability studies with respect to the source condition started with the seminal paper of Burger and Osher [7] who show that the source condition implies stability of the solution according to the Bregman divergence associated to J. In particular, if the source condition is satisfied for $f \in L^2(\mathbb{R}^2) \cap BV(\mathbb{R}^2)$, so that there exists $v = -\operatorname{div} z \in \partial J(f) \neq \emptyset$, then one has support stability in each region U where $\sup_{x \in U} |z(x)| < 1$. Formally, we say that z is *non-degenerate* on U and smaller values of $\sup_{x \in U} |z(x)|$ will allow for stronger stability bounds. More precisely, it was proved in [7] that for $U \subset \mathbb{R}^2$ and $\delta > 0$ such that $\sup_{x \in U} |z(x)| < 1 - \delta$, the solution $u_{\lambda,w}$ to $\mathcal{P}_\lambda(f + w)$ satisfies

$$(1-\delta) \int_U |Du_{\lambda,w}| \leqslant \frac{\|w\|_{L^2}^2}{2\lambda} + \frac{\lambda \|v\|_{L^2}^2}{2} + \|w\|_{L^2} \|v\|_{L^2}.$$

We remark however that this result does not lead to a precise geometric characterization of the TV regularized solutions. In this paper, we shall describe when $u_{\lambda,w}$ is constant on U and hence, our analysis can be seen as a generalization and refinement of this source condition approach.

2. Duality and the minimal norm certificate

We assume for the remainder of this paper that $f \in L^2(\mathbb{R}^2)$ and that the source condition holds. Instead of studying the properties of any element in $\partial J(f)$ as done in [7], we consider the minimal norm certificate:

$$v_0 \overset{\text{def.}}{=} \operatorname{argmin} \left\{ \|v\|_{L^2} \; ; \; v \in \partial J(f) \right\}. \tag{2}$$

The study of v_0 will lead to an understanding of the geometric properties of $u_{\lambda,w}$ in the *low noise regime*, where $\lambda_0 > 0$, $\alpha_0 > 0$, $(\lambda, w) \in D_{\lambda_0,\alpha_0}$ and

$$D_{\lambda_0,\alpha_0} \overset{\text{def.}}{=} \left\{ (\lambda, w) \in \mathbb{R}_+ \times \mathrm{L}^2\left(\mathbb{R}^2\right) \; ; \; 0 \leqslant \lambda \leqslant \lambda_0 \quad \text{and} \quad \|w\|_{L^2} \leqslant \alpha_0 \lambda \right\}. \tag{3}$$

This particular element v_0 naturally arises from the dual problems associated with $(\mathcal{P}_\lambda(f))$, which we now describe. The Fenchel-Rockafellar dual problem of $(\mathcal{P}_\lambda(f))$ is

$$\sup_{v \in \partial J(0)} \langle f, v \rangle - \frac{1}{2\lambda} \|v\|_{L^2}^2, \tag{$\mathcal{D}'_\lambda(f)$}$$

$$\text{or equivalently} \qquad \inf_{v \in \partial J(0)} \|\frac{f}{\lambda} - v\|_{L^2}^2 \tag{$\mathcal{D}_\lambda(f)$}$$

We remark that since the solution $v_{\lambda,w}$ to $(\mathcal{D}_\lambda(f + w))$ is simply the projection of f/λ onto a convex set, $v_{\lambda,w}$ exists and is unique. Furthermore, there is strong duality.

The limit of $(\mathcal{P}_\lambda(f))$ as $\lambda \to 0^+$ is the trivial problem

$$\min_{u \in L^2(\mathbb{R}^2) \cap BV(\mathbb{R}^2)} J(u) \quad \text{s.t.} \quad u = f, \tag{$\mathcal{P}_0(f)$}$$

having $u = f$ as the solution. The dual is

$$\sup_{v \in \partial J(0)} \langle f, v \rangle, \tag{$\mathcal{D}_0(f)$}$$

with solutions $\partial J(f)$. Again, there is strong duality, although there is no guarantee of uniqueness or existence of solutions to $(\mathcal{D}_0(f))$ as it is possible that $\partial J(f) = \emptyset$. Note that since the dual solution $v_{\lambda,w}$ of $(\mathcal{D}_\lambda(f+w))$ is the projection of $(f+w)/\lambda$ onto a convex set, the non-expansiveness of the projection yields

$$\forall (\lambda, w) \in \mathbb{R}_+^* \times \mathrm{L}^2\left(\mathbb{R}^2\right), \quad \|v_{\lambda,0} - v_{\lambda,w}\|_{L^2} \leqslant \frac{\|w\|_{L^2}}{\lambda} \leqslant \alpha_0.$$

As a result, the properties of $v_{\lambda,w}$ are governed by those of $v_{\lambda,0}$, and it turns out that the properties of $v_{\lambda,0}$ are governed, in the low noise regime, by those of the minimal L^2 norm solution to $(\mathcal{D}_0(f))$, as the next result hints.

Journal of Physics: Conference Series **756** (2016) 012007 doi:10.1088/1742-6596/756/1/012007

Proposition 1. *[2] If the source condition holds, then* $\lim_{\lambda\to 0}\|v_{\lambda,0}-v_0\|_{L^2}=0$ *where* v_0*, the minimal norm certificate is defined as in* (2).

The main point in studying the dual problems is that their solutions $v_{\lambda,w}$ are related to the primal solutions $u_{\lambda,w}$ by the extremality relations

$$v_{\lambda,w}\in\partial J(u_{\lambda,w}),\qquad v_{\lambda,w}=\frac{1}{\lambda}(f-u_{\lambda,w}),$$

which makes possible the analysis of the support of $Du_{\lambda,w}$, as described by the following proposition. The intuition behind our study of the minimal norm certificate is that when $v_{\lambda,w}$ is close to v_0, the support of $Du_{\lambda,w}$ will be governed by v_0.

Proposition 2 ([2])**.** *Let* $f,v\in L^2(\mathbb{R}^2)$*. For* $t>0$*, let* $F^{(t)}\overset{\text{def.}}{=}\{f\geqslant t\}$ *and for* $t<0$*, let* $F^{(t)}\overset{\text{def.}}{=}\{f\leqslant t\}$*. The following hold.*

(i) $v\in\partial J(f)$ *if and only if* $v\in\partial J(0)$ *and the level sets of* f *satisfy*

$$\forall t>0,\quad P(F^{(t)})=\int_{F^{(t)}}v,\quad \forall t<0,\quad P(F^{(t)})=-\int_{F^{(t)}}v.$$

(ii) $\operatorname{Supp}(Df)=\overline{\bigcup\{\partial^*F^{(t)}\,;\,t\in\mathbb{R}\setminus\{0\}\}}\subseteq\overline{\bigcup\{\partial^*F\,;\,|F|<+\infty\quad\textit{and}\quad P(F)=\pm\int_F v\}}.$

Examples of minimal norm certificates

- If C is a calibrable set, then $v_0=h_C\mathbb{1}_C$.
- If C is a convex set such that ∂C is of class $C^{1,1}$ and $\operatorname{ess\,sup}_p \kappa_{\partial C}(p)\leqslant c$ for some $c>0$, then there exists some R^* such that C_{R^*} (as defined in Section 1.1) is the maximal calibrable set inside C. In this case, $v_0(x)=1/r$ for $x\in\partial C_r$ and $r\in[0,R^*]$ and $v_0(x)=0$ otherwise.

3. Support stability

In this section, we first define the extended support and state our main result.

3.1. *The extended support*

Motivated by Proposition 2, the extended support of $f\in L^2(\mathbb{R}^2)$ with respect to its minimal norm certificate v_0 is defined as follows.

Definition 1.

$$\operatorname{Ext}(Df)\overset{\text{def.}}{=}\overline{\bigcup\{\operatorname{supp}(Dg)\,;\,v_0\in\partial J(g)\}}=\overline{\bigcup\left\{\partial^*E\,;\,|E|<\infty,\ \pm\int_E v_0=P(E)\right\}}.$$

Examples of the extended support

- If C is a calibrable set, then $\operatorname{Ext}(D\mathbb{1}_C)=\partial C$.
- If C is a smooth convex set, then $\operatorname{Ext}(D\mathbb{1}_C)=C\setminus\operatorname{int}(C_{R^*})$, where C_{R^*} is the maximal calibrable set inside C.

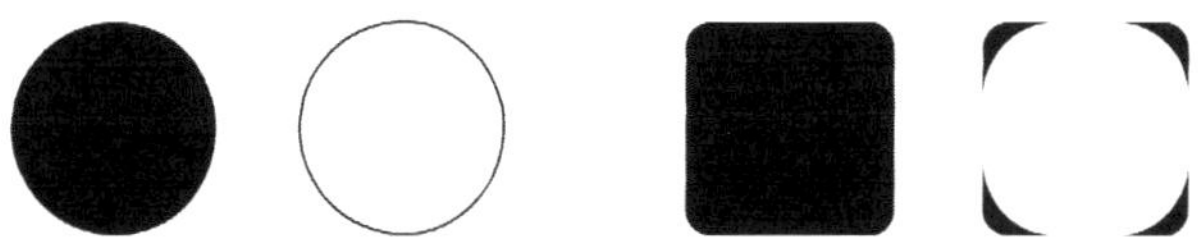

$\mathbb{1}_A$ $\operatorname{Ext}(D\mathbb{1}_A)$ $\mathbb{1}_B$ $\operatorname{Ext}(D\mathbb{1}_B)$

3.2. Main result

We are now ready to present our main result which relates support stability to the extended support.

Theorem 1. *[2] As* $\lambda \to 0$ *and* $\|w\|_{L^2}/\lambda \to 0$,

$$\sup_{x \in \operatorname{supp}(Df)} \operatorname{dist}(x, \operatorname{supp}(Du_{\lambda,w})) \to 0, \qquad \sup_{x \in \operatorname{supp}(Du_{\lambda,w})} \operatorname{dist}(x, \operatorname{Ext} Df) \to 0,$$

Furthermore, by denoting by $U^{(t)}_{\lambda,w}$ *the level sets of* $u_{\lambda,w}$, *for almost every* $t \in \mathbb{R}$, *as* $\lambda \to 0$ *and* $\|w\|_{L^2}/\lambda \to 0$,

$$\left|U^{(t)}_{\lambda,w} \Delta F^{(t)}\right| \to 0, \quad \text{and} \quad \lim_{n \to +\infty} \partial U^{(t)}_{\lambda,w} \to \partial F^{(t)},$$

and the last equality holds in the sense of Hausdorff convergence.

Remark 1. The first part of the theorem above implies that for all $r > 0$, there exists λ_r and α_r such that for all $(\lambda, r) \in D_{\lambda_r,\alpha_r}$, $\operatorname{Supp}(Du_{\lambda,w}) \subseteq T_r \overset{\text{def.}}{=} \left\{x \in \mathbb{R}^2 \,;\, \operatorname{dist}(x, \operatorname{Ext}(Df)) \leqslant r\right\}$, and $\operatorname{Supp}(Df) \subseteq \left\{x \in \mathbb{R}^2 \,;\, \operatorname{dist}(x, \operatorname{Supp}(Du_{\lambda,w})) \leqslant r\right\}$. Thus, one can always define an arbitrarily small tube T_r around the extended support such that $u_{\lambda,w}$ is constant outside T_r for all (λ, w) in some low noise regime.

3.3. Rate of convergence

By definition of v_0, there exists some vector field $z_0 \in L^\infty(\mathbb{R}^2; \mathbb{R}^2)$ such that $v_0 = -\operatorname{div} z_0$ and $\|z_0\|_{L^\infty} \leqslant 1$. This vector field is not unique and our main result is not dependent on its behaviour. However, our next result shows that when the non-degeneracy of z_0 is known (as described in Section 1.1), we can make explicit the relationship between the width of the tube T_r, the decay of $\|v_{\lambda,w} - v_{0,0}\|_{L^2}$ and the nondegeneracy of z_0. Assume now that for all $r > 0$, there exists $\delta_r > 0$ such that

$$1 - \delta_r = \operatorname*{ess\,sup}_{x \notin T_r} |z_0(x)| \,. \tag{4}$$

Theorem 2. *[2] If* $\lambda > 0$ *and* $w \in L^2(\mathbb{R}^2)$ *is such that* $\|v_{\lambda,w} - v_0\|_{L^2} \leqslant \delta_{r/2} \min\{r/(2C),\ 2\sqrt{\pi}\}$, *then* $\operatorname{Supp}(Du_{\lambda,w}) \subset T_r$.

Remark 2. Let us remark on the condition (4). As a result of the constructions described in [8], if C is a calibrable set, then the vector field z_0 of minimal norm certificate associated with $\mathbb{1}_C$ is such that every $K \subset C$ compact, $\sup_K |z_0| < 1$. Furthermore, we can show [2] that for a convex set C with $\mathscr{C}^2$ boundary, $|z_0(s, r)| \lesssim \min\{1 - \kappa_{\partial C}(s) r,\ 1 - r^2\}$ outside C.

Example: If $f = \mathbb{1}_{B(0,R)}$ and $\|w\|_{L^2} \leqslant \lambda r^2/(8R^2)$, then $\operatorname{supp}(Du_{\lambda,w}) \subset T_r$.

4. Numerical examples of the extended support

Figure 1 presents some numerical examples of denoised solutions, the extended support and the vector field z_0 associated with several indicator functions of sets. These examples are computed by solving the dual problem $(\mathcal{D}_\lambda(f))$ using the projected gradient descent algorithm of [9].

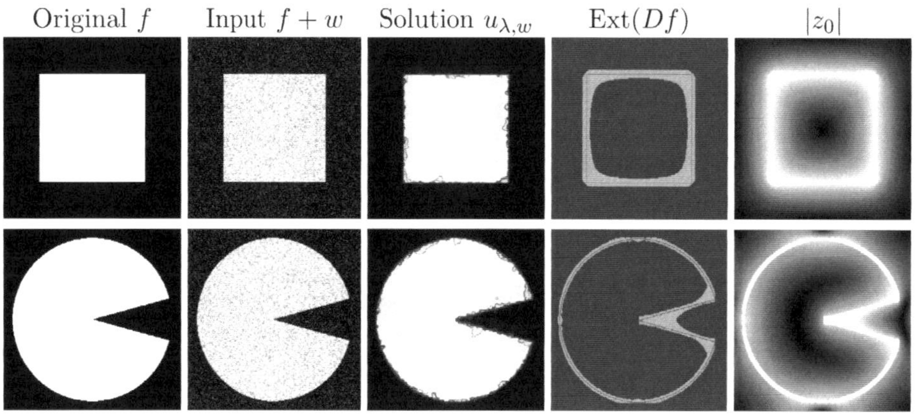

Figure 1: The level lines of each denoised solution are shown in blue. The extended support for each indicator function is shown in gray, and the absolute value of the associated vector field is shown such that white corresponds to one and black corresponds to zero. Observe that the decay of the vector field is linked to the curvature of the boundary as suggested by our remark in the previous section.

5. Conclusion

Our main result shows that although the support of TV regularized solutions are in general not stable and staircasing effects can be introduced, in the low noise regime, the instabilities are confined to a small neighbourhood of the extended support. Furthermore, for the indicator set of a calibrable set C, the support of TV regularized solutions cluster around ∂C. Finally, although the situation where the source condition is not satisfied was not discussed in the present paper, our approach can be extended to characterize the support stability for various functions without the source condition, such as $\mathbb{1}_{[0,1]^2}$ (see [2] for details).

Acknowledgements

V.D and C.P. acknowledge support from the CNRS project CAVALIERI. G.P. acknowledges support from the European Research Council (project SIGMA-Vision). C.P. acknowledges support from the FSMP.

References

[1] Rudin L I, Osher S and Fatemi E 1992 *Physica D: Nonlinear Phenomena* **60** 259–268

[2] Chambolle A, Duval V, Peyré G and Poon C 2016 *Preprint* URL `http://arxiv.org/abs/1602.00087`

[3] Caselles V, Chambolle A and Novaga M 2007 *Multiscale Modeling and Simulation* **6** 879–894

[4] Valkonen T 2015 *SIAM J. Math. Analysis* **47** 2587–2629

[5] Alter F, Caselles V and Chambolle A 2005 *Math. Ann.* **7** 29–53

[6] Scherzer O, Grasmair M, Grossauer H, Haltmeier M and Lenzen F 2009 *Variational Methods in Imaging* 1st ed Applied Mathematical Sciences (Springer) ISBN 0387309314

[7] Burger M and Osher S 2004 *Inverse problems* **20** 1411

[8] Giusti E 1978 *Invent. Math.* **46** 111–137

[9] Chambolle A 2004 *Journal of Mathematical imaging and vision* **20** 89–97

Journal of Physics: Conference Series **756** (2016) 012008 doi:10.1088/1742-6596/756/1/012008

A computational method for the inversion of wide-band GPR measurements

M Salucci[1,2], L Tenuti[2], L Poli[2], G Oliveri[1,2] and A Massa[1,2]
[1] ELEDIA@L2S - Laboratoire des Signaux et Systèmes, UMR8506 (CNRS - CS - UPS), 3 rue Joliot-Curie, 91192 Gif-sur-Yvette, France
[2] ELEDIA@UniTN - University of Trento, Via Sommarive 9, I-38123 Trento, Italy
E-mail: andrea.massa@l2s.centralesupelec.fr

Abstract. An innovative method for the inversion of ground penetrating radar (GPR) measurements is presented. The proposed inverse scattering (IS) approach is based on the exploitation of wide-band data according to a multi-frequency (MF) strategy, and integrates a customized particle swarm optimizer (PSO) within the iterative multi-scaling approach (IMSA) to counteract the high non-linearity of the optimized cost function. If from the one hand the IMSA provides a reduction of the ratio between problem unknowns and informative data, on the other hand the stochastic nature of the PSO solver allows to "escape" from the high density of false solutions of the MF-IS subsurface problem. A set of representative numerical results verifies the effectiveness of the developed approach, as well as its superiority with respect to a deterministic implementation.

1. Introduction

In the last decades, there has been a growing interest in developing non-destructive technologies aimed at retrieving an easy-to-interpret image of an inaccessible buried domain [1],[2]. Within this context, the use of ground penetrating radar (GPR) as probing instrument gained particular attention given its robustness and versatility [1]-[7]. Nevertheless, the need for effective computational methods for the inversion of GPR data is still an open issue. Classical microwave tomography techniques introduced for the free-space scenario seem a promising candidate for solving the subsurface inverse scattering (IS) problem. However, suitable extensions of such techniques are needed in order to take into account for the increased complexity of the buried configuration, as well as for the limited amount of information that can be collected by field probes placed only above the interface [2].

When developing GPR microwave techniques, it is worth remarking that an intrinsic frequency diversity is available in the collected measurements thanks to the exploitation of wide-band time-domain pulses, and such a diversity should be regarded as an additional source of information able to mitigate the *ill-posedness* of the subsurface IS problem [3]. As a matter of fact, several approaches able to deal with such wide-band data have been recently proposed, mainly subdivided in (*i*) frequency-hopping (FH) [3],[7] and (*ii*) multi-frequency (MF) [4],[8],[9] strategies. If from the one hand FH-based methods process different samples of the collected GPR spectrum in a cascaded fashion [3], MF-IS techniques exploit the whole data within a single inversion step. According to the reference literature, MF strategies could in principle provide better results than FH-based methods since the joint exploitation of multiple frequencies enlarges the essential dimension of the available data [8]. However, they suffer from an increased complexity of the cost function that has to be minimized in terms of occurrence of local minima, since a larger number of unknowns has to be retrieved within a single optimization. In such a framework, multi-resolution techniques [3],[9] can be introduced in the inversion scheme to limit the ratio between unknowns and uncorrelated data, by adaptively refining the resolution only within those regions where the targets have been detected. Unfortunately, such approaches only partially solve the problem as they cannot completely prevent local search algorithms from being trapped into false solutions (mathematically represented by local minima of the MF cost function to optimize).

Given all the previous considerations, this work presents an innovative computational method for the inversion of GPR measurements. Such a methodology is based on the integration of the iterative multi-scaling approach (IMSA [9]) and a MF stochastic solver based on the particle swarm optimizer (PSO) [10]-[12]. As shown by a set of representative numerical comparisons, the MF-IMSA-PSO is not only able to provide accurate and robust reconstructions of the buried scenario, but it also significantly overcomes its deterministic counterpart based on a conjugate gradient (CG)-based solver thanks to its increased global convergence capability.

2. Problem formulation and solution approach

Let us consider a standard two-dimensional (2D) GPR prospecting configuration. The upper half-space ($y>0$) is free-space (ε_0 and σ_0), while the lower half-space ($y<0$) is occupied by a background medium with complex permittivity equal to $\varepsilon_{eqb}(f)=\varepsilon_{rb}\varepsilon_0 - j\sigma_b/2\pi f$. A buried investigation domain D_{inv} contains an unknown scatterer having relative permittivity and conductivity equal to ε_r^{obj} and σ^{obj}, respectively. D_{inv} is imaged by means of V ideal z-oriented infinite line sources located at constant height above the interface and radiating a wide-band pulse with 3 dB bandwidth covering the interval $[f_{\min}, f_{\max}]$. Under these hypothesis, the time-domain scattered field collected by M probes co-located with the sources [$\Psi_s^{(v)}\left(\underline{r}_m^{(v)},t\right)$, $v=1,...,V$, $m=1,...,M$] is extracted by muting the first temporal interval of the measured total field [3]. Then, after transforming Ψ_s into the frequency-domain, a set of Q frequency samples in the $[f_{\min}, f_{\max}]$ band are extracted from the computed spectrum

$$E_s^{(v)}\left(\underline{r}_m^{(v)}, f_q\right)=F\left\{\Psi_s^{(v)}\left(\underline{r}_m^{(v)},t\right)\right\}\Big|_{f=f_q} \qquad v=1,...,V \quad m=1,...,M \quad q=1,...,Q \tag{1}$$

where $F\{a(t)\}=\int_{-\infty}^{+\infty} a(t)\exp(-j2\pi f t)dt$. The q-th sample of E_s is then linked to the unknown total field E_t and to the electromagnetic properties of D_{inv} by means of the non-linear integral *data* equation

$$E_s^{(v)}\left(\underline{r}_m^{(v)}, f_q\right)=\int_{D_{inv}} G_q^{ext}\left(\underline{r}_m^{(v)},\underline{r}'\right)E_t^{(v)}\left(\underline{r}', f_q\right)\tau\left(\underline{r}', f_q\right)d\underline{r}' \qquad v=1,...,V \quad m=1,...,M \quad q=1,...,Q \tag{2}$$

where G_q^{ext} is the *external* Green's function for the half-space scenario and τ is the unknown contrast

$$\tau\left(\underline{r}, f_q\right)=\frac{\varepsilon_{eq}\left(\underline{r}, f_q\right)-\varepsilon_{eqb}\left(f_q\right)}{\varepsilon_0} \qquad q=1,...,Q \quad \underline{r}\in D_{inv} \tag{3}$$

with $\varepsilon_{eq}\left(\underline{r}, f_q\right)=\varepsilon_r\left(\underline{r}\right)\varepsilon_0 - j\sigma\left(\underline{r}\right)/2\pi f_q$. Similarly, the incident field for $\underline{r}\in D_{inv}$ is linked to the problem unknowns by means of the *state* equation

$$E_i^{(v)}\left(\underline{r}, f_q\right)=E_t^{(v)}\left(\underline{r}, f_q\right)-\int_{D_{inv}} G_q^{\text{int}}\left(\underline{r},\underline{r}'\right)E_t^{(v)}\left(\underline{r}', f_q\right)\tau\left(\underline{r}', f_q\right)d\underline{r}' \tag{4}$$

$$v=1,...,V \quad m=1,...,M \quad q=1,...,Q$$

where G_q^{int} stands for the *internal* Green's function. Under these hypothesis, in the MF framework the goal of the inverse problem is to find an estimate of E_t and τ inside D_{inv} by finding the global minimum of the following cost function

$$\Phi = \frac{\sum_{m,v,q} \left| E_s^{(v)}\left(\underline{r}_m^{(v)}, f_q\right) - \tilde{E}_s^{(v)}\left(\underline{r}_m^{(v)}, f_q\right) \right|^2}{\sum_{m,v,q} \left| E_s^{(v)}\left(\underline{r}_m^{(v)}, f_q\right) \right|^2} + \frac{\sum_{n,v,q} \left| E_i^{(v)}\left(\underline{r}_n, f_q\right) - \tilde{E}_i^{(v)}\left(\underline{r}_n, f_q\right) \right|^2}{\sum_{n,v,q} \left| E_i^{(v)}\left(\underline{r}_n, f_q\right) \right|^2} \tag{5}$$

where $\tilde{E}_s^{(v)}$ and $\tilde{E}_i^{(v)}$ are the retrieved scattered and incident fields computed through (2) and (4), while N is the number of square basis functions used to discretize D_{inv}. The solution of such a problem is obtained by integrating a customized MF particle swarm optimizer (MF-PSO) within an iterative multi-scaling approach (IMSA) made of S multi-zooming steps as follows:

1. *IMSA Initialization (s = 0).* Subdivide D_{inv} into N square sub-domains;
2. *Low-Resolution Reconstruction.* Minimize Φ in I iterations by means of the MF-PSO:
   a. *Swarm Initialization (i = 0).* Generate a swarm of P particles by initializing their positions as random perturbations of the empty-background configuration and considering random velocities. Towards this end, the following relationship between the contrast at each frequency and at a reference frequency $\overline{f}$ is exploited

$$\tau\left(\underline{r}_n, f_q\right) = \mathrm{Re}\left\{\tau\left(\underline{r}_n, \overline{f}\right)\right\} + j\frac{\overline{f}}{f_q}\mathrm{Im}\left\{\tau\left(\underline{r}_n, \overline{f}\right)\right\} \quad n = 1,\ldots,N \quad q = 1,\ldots,Q; \tag{6}$$

   b. *Swarm Evaluation.* Compute Φ for each particle and update the global and personal best positions;
   c. *Swarm Updating.* Update the velocity and the position of each particle to make the swarm evolve towards the next iteration ($i = i+1$);
   d. *PSO Termination.* Stop when a maximum number of iterations I has been reached;
3. *Multi-Resolution Loop (s=2,...,S).* Identify the region of interest $D^{(s)} \subset D^{(s-1)}$ ($D^{(1)} = D_{inv}$) and discretize it into N cells.
   a. *High-Resolution Reconstruction.* Exploit the information collected at the $(s-1)$-th step to initialize the swarm and minimize the cost function through the MF-PSO;
   b. *IMSA Termination.* End the multi-zooming loop once a suitable stopping criterion is met ($s = S$).

3. Numerical results

The goal of this section is to numerically assess the effectiveness and the robustness of the MF-IMSA-PSO when blurring the collected time-domain total field with the desired level of signal-to-noise ratio (SNR). Moreover, its performances will be compared to those of the MF-IMSA-CG, a deterministic implementation exploiting a CG-based MF-IS solver.

3.1. Performance comparison

Let us consider the target in Fig. 1(*a*) (*L-Shaped* profile, $\varepsilon_r^{obj} = 5$ and $\sigma^{obj} = 0.001$ S/m). A square D_{inv} (side 0.8 m) is embedded in a lossy soil with $\varepsilon_{rb} = 4$ and $\sigma_b = 0.001$ S/m, while $V = 10$ sources and $M = 9$ probes are used to image it, by simulating the buried scenario in time-domain by means of the *GPRMax2D* simulator [13]. $Q = 5$ equally-spaced frequency samples are extracted from the collected GPR spectrum in the $[200, 600]$ MHz band. Fig. 1(*b*) shows the retrieved profile (at $\overline{f} = 400$

MHz) by the MF-IMSA-PSO when a significant amount of noise is added to the collected *radargram* ($SNR = 20$ dB). As it can be easily observed, the target is correctly localized and shaped, with a good estimation of the contrast. Moreover, the MF-IMSA-PSO significantly outperforms the MF-IMSA-CG, this latter providing a strong under-estimation of the contrast [Fig. 1(*c*) vs. Fig. 1(*b*)].

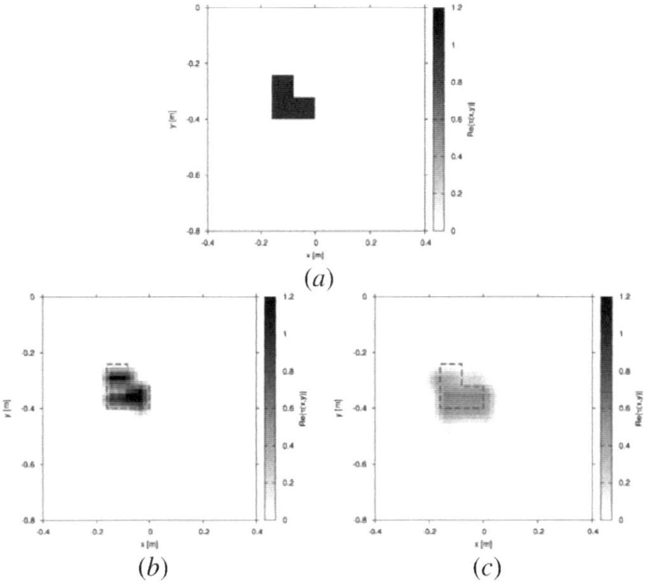

Figure 1. *L-shaped scatterer* - (*a*) Actual dielectric profile and retrieved profiles by (*b*) MF-IMSA-PSO and (*c*) MF-IMSA-CG (*SNR* = 20 dB).

Such a superiority is even more evident for a *I-Shaped* profile [$\varepsilon_r^{obj} = 5$ and $\sigma^{obj} = 0.001$ S/m - Fig. 2(*a*)]. As a matter of fact, the MF-IMSA-CG fails in correctly identifying the support of the target [Fig. 2(*c*)], while a very accurate estimation is provided by the proposed approach [Fig. 2(*b*)].

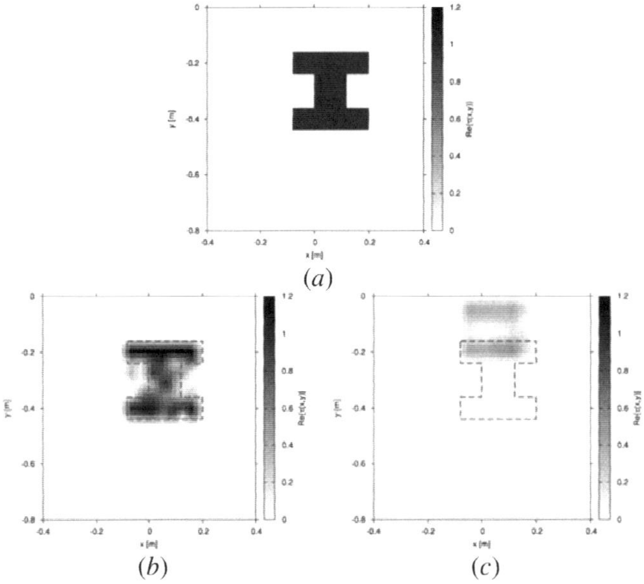

Figure 2. *I-shaped scatterer* - (*a*) Actual dielectric profile and retrieved profiles by (*b*) MF-IMSA-PSO and (*c*) MF-IMSA-CG (*SNR* = 20 dB).

Journal of Physics: Conference Series **756** (2016) 012008 doi:10.1088/1742-6596/756/1/012008

3.2. Circular target at different depths

In order to further investigate the robustness and the effectiveness of the MF-IMSA-PSO, let us consider the retrieval of a circular target ($\varepsilon_r^{obj} = 5$ and $\sigma^{obj} = 0.001$ S/m) buried at different depths inside D_{inv} [$d^{obj} = 0.16$ m - Figs. 3(*a*),3(*b*), $d^{obj} = 0.4$ m - Figs. 3(*c*),3(*d*) and $d^{obj} = 0.64$ m - Figs. 3(*e*),3(*f*)]. As it can be seen from the reported reconstructions on the right column of Fig. 3, the developed methodology is able to correctly locate and shape the target is all cases, even if an unavoidable degradation of the accuracy is observed when increasing the depth of the scatterer inside the lossy background [Fig. 3(*f*) vs. Fig. 3(*b*)].

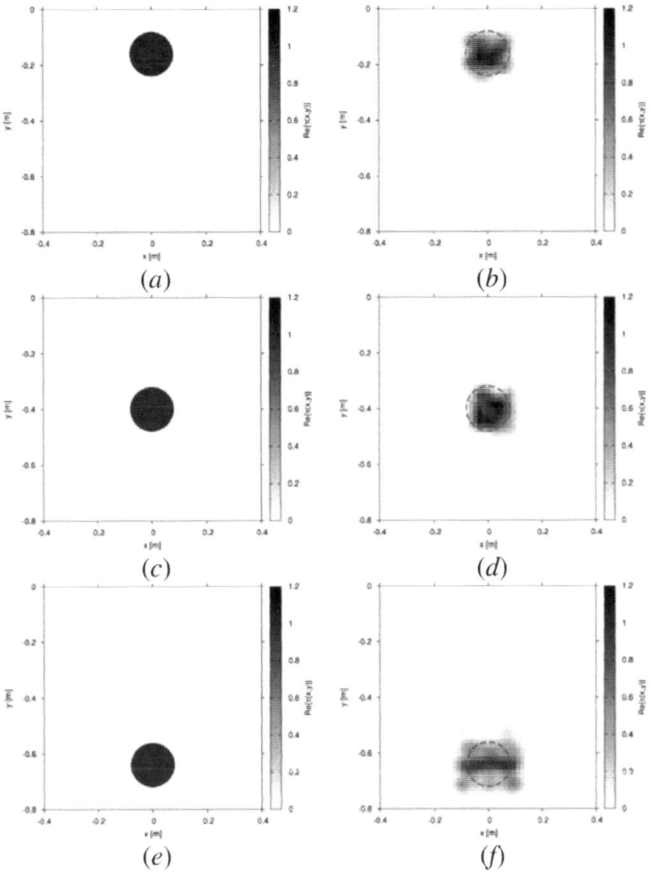

Figure 3. *Circular scatterer* - (*a*)(*c*)(*e*) Actual profiles and (*b*)(*d*)(*f*) MF-IMSA-PSO reconstructions when (*a*)(*b*) $d^{obj} = 0.16$ m, (*c*)(*d*) $d^{obj} = 0.4$ m and (*e*)(*f*) $d^{obj} = 0.64$ m (*SNR* = 40 dB).

4. Conclusions

A new computational method for the inversion of multi-frequency GPR data has been proposed. The MF-IMSA-PSO benefits from a reduction of the ratio between problem unknowns and informative data thanks to the IMSA, while provides robust minimizations of the MF cost function thanks to the *hill-climbing* properties of the employed stochastic optimizer. A set of numerical examples has validated the proposed approach, that showed its superiority over a deterministic CG-based implementation.

References

[1] Lesselier D and Habashy T 2000 Special issue on electromagnetic imaging and inversion on the Earth's subsurface *Inverse Prob.* **16**

[2] Chen C.-C., Johnson J. T., Sato M. and Yarovoy A. G. 2007 Special issue on subsurface sensing using ground-penetrating radar *IEEE Trans. Geosci. Remote Sens.* **45**

[3] Salucci M, Oliveri G and Massa A 2015 GPR prospecting through an inverse scattering frequency-hopping multi-focusing approach *IEEE Trans. Geosci. Remote Sens.* **53** 6573-6592

[4] Persico R, Soldovieri F and Leone G 2006 A microwave tomographic imaging approach for multibistatic configuration: the choice of the frequency step *IEEE Trans. Instrum. Meas.* **55** 1926-1934

[5] Salucci M, Oliveri G, Randazzo A, Pastorino M and Massa A 2014 Electromagnetic subsurface prospecting by a multifocusing inexact Newton method within the second-order Born approximation *J. Opt. Soc. Am. A* **31** 1167-1179

[6] Salucci M, Oliveri G, Randazzo A, Pastorino M and Massa A 2014 Electromagnetic subsurface prospecting by a fully nonlinear multi-focusing inexact Newton method *J. Opt. Soc. Am. A* **31** 2618-2629

[7] Moriyama T, Salucci M, Oliveri G, Tenuti L, Rocca P and Massa A 2014 A Multi-scaling deterministic imaging for GPR survey *Proc. 2014 IEEE Antenna Conf. Ant. Meas. Appl. (Antibes Juan-les-Pins)* 1-3

[8] Bucci O M, Crocco L, Isernia T and Pascazio V 2000 Inverse scattering problems with multifrequency data: reconstruction capabilities and solution strategies *IEEE Trans. Geosci. Remote Sens.* **38** 1749-1756

[9] Franceschini D, Donelli M, Azaro R and Massa A 2007 Dealing with multi-frequency scattering data through the IMSA *IEEE Trans. Antennas Propag.* **55** 2412-2417

[10] Rocca P, Benedetti M, Donelli M, Franceschini D and Massa A 2009 Evolutionary optimization as applied to inverse problems *Inverse Problems* **25** 1-41

[11] Donelli M and Massa A 2005 Computational approach based on a particle swarm optimizer for microwave imaging of two-dimensional dielectric scatterers *IEEE Trans. Microw. Theory Tech.* **53** 1761-1776

[12] Donelli M, Franceschini D, Rocca P and Massa A 2009 Three-dimensional microwave imaging problems solved through an efficient multiscaling particle swarm optimization *IEEE Trans. Geosci. Remote Sens.* **47** 1467-1481

[13] Giannopoulos A 2005 Modelling ground penetrating radar by GprMax *Construct. Build. Mater.* **19** 755 -762

Acknowledgment

This work has been partially supported by the SIRENA project (2014-2017) funded by DIGITEO (France) under the "Call for Chairs 2014" and benefited from the networking activities carried out within the EU funded COST Action TU1208 "Civil Engineering Applications of Ground Penetrating Radar".

The ROI CT problem: a shearlet-based regularization approach

T A Bubba[1], F Porta[2], G Zanghirati[2] and S Bonettini[2]

[1] Dipartimento di Scienze Fisiche, Informatiche e Matematiche, Università di Modena e Reggio Emilia, Via Campi 213/b, 41125 Modena, Italy

[2] Dipartimento di Matematica e Informatica, Università di Ferrara, Via Saragat 1, 44122 Ferrara, Italy

E-mail: tatiana.bubba@unimore.it, federica.porta@unife.it, g.zanghirati@unife.it, silvia.bonettini@unife.it

Abstract. The possibility to significantly reduce the X-ray radiation dose and shorten the scanning time is particularly appealing, especially for the medical imaging community. Region-of-interest Computed Tomography (ROI CT) has this potential and, for this reason, is currently receiving increasing attention. Due to the truncation of projection images, ROI CT is a rather challenging problem. Indeed, the ROI reconstruction problem is severely ill-posed in general and naive local reconstruction algorithms tend to be very unstable. To obtain a stable and reliable reconstruction, under suitable noise circumstances, we formulate the ROI CT problem as a convex optimization problem with a regularization term based on shearlets, and possibly nonsmooth. For the solution, we propose and analyze an iterative approach based on the variable metric inexact line-search algorithm (VMILA). The reconstruction performance of VMILA is compared against different regularization conditions, in the case of fan-beam CT simulated data. The numerical tests show that our approach is insensitive to the location of the ROI and remains very stable also when the ROI size is rather small.

1. Introduction

Region-of-interest Computed Tomography (ROI CT) is an X-ray based incomplete data imaging acquisition modality [1]. Since X-ray radiation exposure comes with health hazards for patients, the possibility to reconstruct only a small ROI using truncated projection data is particularly appealing, especially in biomedical application, due to its potential to lower the X-ray radiation dose and reduce the scanning time. However, reconstructing a density function from its projections is an ill-posed problem, with the ill-posedness becoming more severe when projections are truncated, as in the case of ROI CT. Therefore, traditional approaches, like Filtered Back-Projection, in general produce unacceptable visual artifacts and are unstable to noise.

To address the problem of ROI reconstruction from truncated projections, a variety of *ad hoc* methods, both analytic and algebraic, were proposed in the last years (see [2] and the references therein), but usually require restrictive hypothesis on the ROI or are rather sensitive to noise.

To overcome these drawbacks, we formulate ROI CT as a convex optimization problem with a regularized objective function, possibly nonsmooth, and based on a very recently introduced multiscale method called shearlets [3]. The use of less recent multiscale methods is not new in CT application, even combined with analytic approaches [4]. The approach we present partially relies on the setup in [5]. However, our approach differs in the objective function, which considers shearlets instead of wavelets, and in the aim, since our goal is to compare the different regularization terms proposed to identify the model which better provides the desired features of the image to reconstruct. Moreover, for the solution of the minimization problems we exploit a very recently proposed algorithm belonging to the class of proximal-gradient techniques, the variable metric inexact line-search algorithm (VMILA) [6], in place of a (split) augmented Lagrangian. VMILA is a proximal-gradient method, which enables the inexact computation of the proximal point defining the descent direction and guarantees the sufficient decrease of the objective function by means of an Armijo–like backtracking procedure. In order to evaluate the effectiveness of the suggested theoretical approaches we conduct a numerical study on simulated data. In particular, we investigate which approach produces the most accurate ROI reconstruction, without any assumption on the ROI size or location.

2. Region-of-interest tomography setup

In the ROI tomography problem, measurements are taken only within a limited region-of-interest (generally, a disk) strictly inside the object support. The goal is to reconstruct the density function inside the ROI only from these truncated data.

This can be accomplished by using the mask function $M(\theta,\tau) = 1_{\mathcal{P}(S)}(\theta,\tau)$ which selects the set $\mathcal{P}(S) = \{(\theta,\tau) \in \mathcal{T} : \ell(\theta,\tau) \cap S \neq \emptyset\}$ of those rays meeting the ROI, where 1_A is the characteristic function of the set A, S the ROI disk, $\ell(\theta,\tau) = \{\mathbf{x} \in \mathbb{R}^2 : \langle \mathbf{x}, \omega_\theta \rangle = \tau\}$ denotes a ray, being $\omega_\theta = (\cos(\theta), \sin(\theta))^T$, and $\mathcal{T} = \{(\theta,\tau) : \theta \in [0, 2\pi), \tau \in \mathbb{R}\}$ the tangent space. Namely, the *truncated sinogram* $y_0(\theta,\tau)$ shall be given by:

$$y_0(\theta,\tau) = M(\theta,\tau)\,\mathcal{R}f(\theta,\tau), \tag{1}$$

where $\mathcal{R}f(\theta,\tau) = \int_{\ell(\theta,\tau)} f(\mathbf{x})\,d\mathbf{x}$ is the Radon transform of the density function $f \in \mathbb{R}^2$ [7]. In the following, we will address as *full sinogram* the Radon projections

$$y(\theta,\tau) = \mathcal{R}f(\theta,\tau). \tag{2}$$

In particular, equation (2) can be used as a global constraint to fix an extrapolation scheme which defines y outside $\mathcal{P}(S)$. Indeed, the following equation

$$y = y_0 + (1 - M)y. \tag{3}$$

suggests to interpret ROI CT as an extrapolation problem, where, given y_0 on $\mathcal{P}(S)$, the goal is to extrapolate this function to the region outside $\mathcal{P}(S)$. Clearly, not *any* extrapolation scheme is suitable to define y outside $\mathcal{P}(S)$ and equation (2) ensures that y is the image of the Radon transform of the same density function $f \in L^1(\mathbb{R}) \cap L^2(\mathbb{R})$. From (2), we derive two conditions:

$$M\mathcal{R}f = M\,y = y_0 \qquad \text{(data fidelity)} \tag{4}$$

$$(1 - M)\mathcal{R}f = (1 - M)\,y \qquad \text{(data consistency)} \tag{5}$$

Notice that the data fidelity equation defines a constraint inside the ROI while the data consistency equation enforces accurate reconstruction inside the ROI.

Equations (4) and (5) alone do not lead to a unique solution. Indeed, the least square solution

$$\hat{f} = \arg\min_f \|M\mathcal{R}f - y_0\|_2^2,$$

is not unique, since, in general, the solution of the ROI CT problem is not guaranteed to be unique [7]. When uniqueness is ensured, the ill-posedness is too severe, due to the truncation of projections. A classical approach to achieve uniqueness is to impose an additional norm condition, according to Tikhonov regularization, and this norm condition can be applied in the image domain or in the (Radon) transform domain. Hence, we define a norm condition in the transform domain by posing:

$$\|f\|_2^2 = \frac{1}{4\pi}\left\|I^{-\frac{1}{2}}\mathcal{R}f\right\|_2^2$$

given that $f=(4\pi)^{-1}\mathcal{R}^*I^{-1}\mathcal{R}f$ holds true [7, p. 11], where $\mathcal{R}^*$ is the adjoint of the operator $\mathcal{R}$ (usually referred to as backprojection operator), and $I^{-1/2}$ is the Riesz potential operator [7, p. 5]. Thus, we are lead to the following optimization problem:

$$\hat{f} = \underset{f}{\operatorname{argmin}}\, \Upsilon_2(f) \qquad \text{s. t.} \quad y = \mathcal{R}f \quad \text{and} \quad My = y_0 \tag{6}$$

with $\Upsilon_2(f) = \|I^{-1/2}\mathcal{R}f\|_2^2$. By analogy, we can modify the optimization problem using the functional $\Upsilon_p(f) = \|I^{-1/2}\mathcal{R}f\|_p^p$ with $1 \leq p \leq 2$. In the following, we will use $p = 1$, that can be interpreted as a sparsity-promoting condition. The discrete counterpart of (6) reads as:

$$\hat{\mathbf{f}} = \underset{\mathbf{f}}{\operatorname{argmin}}\, \|\mathbf{\Phi}\mathbf{W}\mathbf{f}\|_p^p \qquad \text{s. t.} \quad \mathbf{y} = \mathbf{W}\mathbf{f} \quad \text{and} \quad \mathbf{M}\mathbf{y} = \mathbf{y}_0 \tag{7}$$

where $\mathbf{W}$ is the forward projection matrix, representing the map from the image domain to the projection domain. It is sized $N_\theta N_{\text{dtc}} \times N^2$, with N_θ number of projection angles, N_{dtc} number of detector cells and N is both the width and the height in pixels of the object to reconstruct. The elements of the matrix $\mathbf{W}$ are computed according to the distance-driven method [8]. The diagonal matrix $\mathbf{M}$, sized $N_\theta N_{\text{dtc}} \times N_\theta N_{\text{dtc}}$, is the mask corresponding to the ROI, whose entries are either 0 or 1. The unknown discrete density function $\mathbf{f}$ is a vector of length N^2, while the full and truncated sinogram $\mathbf{y}$ and $\mathbf{y}_0$ are vectors of length $N_\theta N_{\text{dtc}}$, all obtained by column-wise stacking the entries of the corresponding matrices. Finally, $\mathbf{\Phi}$ is a discrete filter *corresponding* to the Riesz potential operator $I^{-1/2}$. Indeed, in place of considering a straightforward matrix discretization of the Riesz potential operator, we approximate it by using a discrete shearlet transform [3], whose corresponding matrix is denoted by $\mathbf{\Phi}$. The underlying idea is to exploit the shearlets optimally sparse approximation properties, which are particularly relevant in CT-like applications, since point-like structures in the image domain map onto sine-shaped curvilinear structures in the projection domain [9]. Formulation (7) is equivalent to the following formulation:

$$\begin{gathered}\hat{\mathbf{f}} = \underset{\mathbf{f}\in\mathbb{R}^{N^2}}{\operatorname{argmin}}\, \mathbf{\Psi}(\mathbf{f}) \qquad \text{where} \\ \mathbf{\Psi}(\mathbf{f}) = \frac{1}{2}\|\mathbf{M}\mathbf{W}\mathbf{f} - \mathbf{y}_0\|_2^2 + \lambda\,\|\mathbf{\Phi}((\mathbf{I}_{N_\theta N_{\text{dtc}}} - \mathbf{M})\mathbf{W}\mathbf{f} + \mathbf{y}_0)\|_p^p + \iota_{\Omega_\mathbf{f}}.\end{gathered} \tag{8}$$

In (8) we considered the L^2-norm error of the data fidelity equation and we incorporated the data consistency information in the regularization term. Here, λ denotes the regularization parameter and $\iota_{\Omega_\mathbf{f}}$ is the indicator function of the feasible region which is defined as $\mathbf{f} \geq 0$ or $0 \leq \mathbf{f} \leq L$, where L is the image maximum pixel intensity and the inequalities are meant component-wise. Notice that each term of the objective function is convex with respect to the unknown. In the following, we compare two different models belonging to (8) and obtained by selecting $p = 2$ or $p = 1$.

3. Numerical Experiments

To demonstrate and validate our approach, we use the synthetic data set known as "modified Shepp-Logan phantom" sized $N \times N$ pixels with $N = 128$, which is available, for instance, in

Table 1. Optimal results for shearlet-based regularization (8) with $p = 1, 2$, and $\lambda = 10^{-4}$.

	$\sigma = 0.05$									$\sigma = 0.1$								
	$\mathbf{r}_{\text{ROI}} = 0.25N$			$\mathbf{r}_{\text{ROI}} = 0.15N$			$\mathbf{r}_{\text{ROI}} = 0.1N$			$\mathbf{r}_{\text{ROI}} = 0.25N$			$\mathbf{r}_{\text{ROI}} = 0.15N$			$\mathbf{r}_{\text{ROI}} = 0.1N$		
	iter	value	sec	iter	value	sec	iter	value	sec	iter	value	sec	iter	value	sec	iter	value	sec
	ROI PSNR									ROI PSNR								
Sm	83	35.86	4.7	109	33.61	6.0	106	32.57	5.8	77	35.49	5.3	111	35.37	7.6	102	37.12	6.9
NSm	83	36.94	15.1	78	41.98	14.2	48	45.43	8.6	48	37.76	10.5	72	40.92	15.9	62	44.64	13.6
	ROI Relative error									ROI Relative error								
Sm	83	0.21	4.7	109	0.57	6.0	106	1.06	5.8	77	0.23	5.3	111	0.46	7.6	102	0.63	6.9
NSm	83	0.19	15.1	78	0.22	14.2	48	0.24	8.6	48	0.17	10.5	72	0.25	15.9	62	0.26	13.6

the Matlab Image Processing toolbox. All the algorithms are implemented in Matlab 8.1.0 and the experiments performed on a dual CPU server, equipped with two 6-cores Intel Xeon X5690 at 3.46GHz, 188 GB DDR3 central RAM memory and up to 12 TB of disk storage.

All the numerical experiments are carried out in the framework of 2D fan-beam geometry. A full angle scan (182 views over 2π) with complete data was simulated. The matrix $\mathbf{\Phi}$ is generated by using the *classical shearlet* decomposition. In details, the number of scales for the shearlet transform is set equal to 4 and the number of directions across the scales is set to (8, 8, 16, 16).

Truncated projection data are obtained by discarding the samples outside the ROI projection $\mathcal{P}(S)$. In particular, we consider concentric ROI disks with decreasing radius $\mathbf{r}_{\text{ROI}}$, placed off-center with respect to the field of view. The results in this paper cover ROIs which are fully inside the object being imaged, according to the hypothesis of the interior tomography problem. Here, we assume that the noise which corrupts the projection data is described by a white Gaussian process, with zero mean and variance σ.

To evaluate the goodness of model (8) by varying the regularization term, we investigated two different formulations: a smooth one with $p = 2$ and a nonsmooth one with $p = 1$. The solution of both the corresponding optimization problems has been addressed by using VMILA. It is a proximal-gradient (or forward-backward) method suitable for minimizing the sum of a differentiable, possibly nonconvex, function plus a convex, possibly nondifferentiable, function. Both the considered formulations for the ROI CT problem are suitable to be faced by such an algorithm. Indeed, in the smooth framework, the differentiable part consists of the first two terms of (8) and the nondifferentiable contribute reduces to the indicator function of the feasible set $\Omega_{\mathbf{f}}$. In the nonsmooth setting, the differentiable term is given by the L^2-norm of the data fidelity and the nondifferentiable one takes into account both $\iota_{\Omega_{\mathbf{f}}}$ and the data consistency regularizer. The VMILA $(k+1)$-th iteration for the minimization of the sum of a smooth function Γ_0 and a nonsmooth one Γ_1 is given by:

$$\mathbf{f}^{(k+1)} = \mathbf{f}^{(k)} + \lambda_k \left\{ \text{prox}_{\alpha_k \Gamma_1}^{D_k^{-1}} \left(\mathbf{f}^{(k)} - \alpha_k D_k \nabla \Gamma_0(\mathbf{f}^{(k)}) \right) - \mathbf{f}^{(k)} \right\}$$

where λ_k is a linesearch parameter, α_k is a positive steplength, D_k is a symmetric and positive definite scaling matrix and $\text{prox}_{\alpha_k \Gamma_1}^{D_k^{-1}}$ denotes the proximal operator associated to Γ_1. VMILA global convergence is ensured by a generalized Armijo-like backtracking procedure to select λ_k. Moreover, $\alpha_k \in [\alpha_{\min}, \alpha_{\max}] \subset \mathbb{R}^+$ and $D_k \in D_\mu$, where $D_\mu = \{D = \text{diag}(d_1, \dots, d_N) \,|\, d_j \in [\mu^{-1}, \mu] \; \forall j\}$ with $\mu > 1$. For both formulations, we select α_k through an adaptive strategy based on the Barzilai-Borwein updating rules [10, 11] and D_k according to a split-gradient idea [12] based on the decomposition of the gradient of Γ_0 into a positive and a negative part.

In many common situations, including ROI CT, a closed-form of the proximal operator of Γ_1 is not known. In this case, VMILA consists in a double loop method, where the inner loop,

Figure 1. Optimal reconstructions of the Shepp-Logan phantom for decreasing radii and $\sigma = 0.05$. First row: smooth formulation. Second row: nonsmooth formulation.

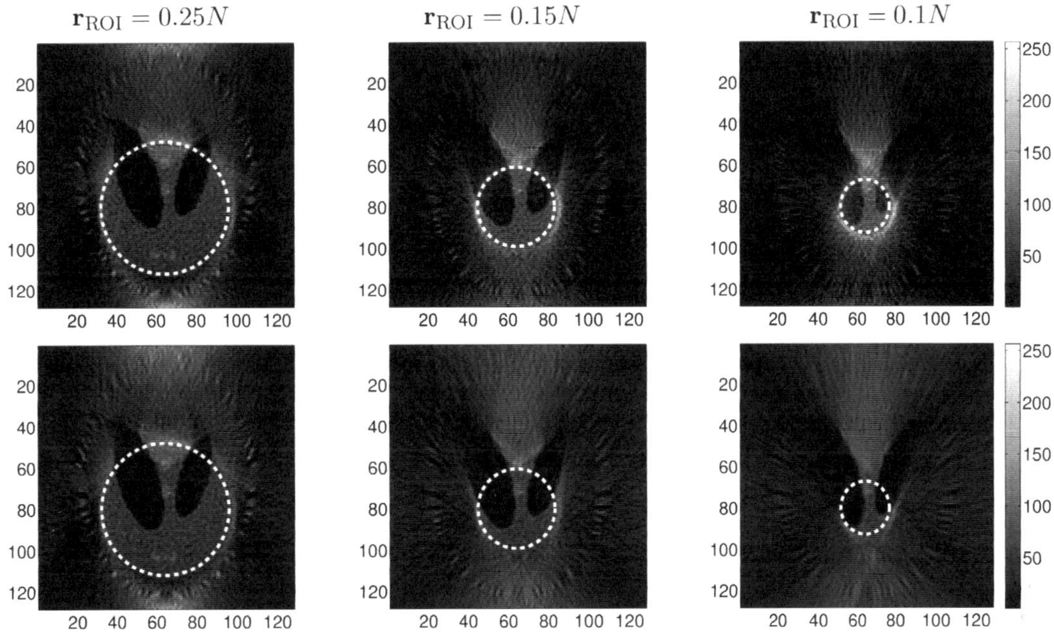

designed to inexactly solve the minimization problem associated to the definition of the proximal operator, is devised with a suitable stopping criterion. For the ROI CT problem, if $p = 1$ we need to compute an approximation of the proximal point which defines the VMILA direction; on the other hand, if $p = 2$, the proximal operator corresponding to the nondifferentiable term of the objective function reduces to the scaled euclidean projection onto the feasible set and therefore no inner loop is required.

We investigated the performance of both the models for regularization parameters $\lambda = 10^{\ell}$, with $\ell = -4, -3, \ldots, 4$. As figures of merit, we use the peak-signal-to-noise ratio (PSNR) and the relative error. We stress that both PSNR and relative error are evaluated inside the ROI only, since ROI CT aims to recover the image only inside the ROI.

Due to room constraint, we summarize in Table 1 only the best results, with respect to the figures of merit, obtained with the shearlet-based formulation for both smooth (Sm) and nonsmooth (NSm) cases, and for two levels of noise, namely $\sigma = 0.05$ (Table 1, left-hand side) and $\sigma = 0.1$ (Table 1, right-hand side). The corresponding images, only for the case $\sigma = 0.05$, are reported in Figure 1. We stress that many additional tests were performed for both formulations, with different level of noise, and also exploiting other regularization approaches, such as smooth total variation (sTV) and early stopping. For instance, for all ROI radii in Table 1 the sTV approach with $\rho = 1$ and $\sigma = 0.05$ yields PSNR $= 37.81, 42.93, 47.25$, relative error equal to 0.17, 0.19, 0.20, and iter $= 107, 145, 172$ in 7.3, 9.7, 11.7 seconds, respectively, while with $\sigma = 0.1$, it yields PSNR $= 39.21, 44.94, 49.69$, relative error equal to 0.14, 0.15, 0.15, and iter $= 114, 137, 181$ in 7.6, 9.4, 12.3 seconds, respectively. Worse performances were obtained for $\rho = 10^{-2}, 10^{-1}, 10$. These results are *slightly* better than the (nonsmooth) shearlets-based approach, but this is clearly dependent on the phantom features (which is piecewise constant) and may not hold for more general data. We expect that, using more realistic sinograms, the contribution of the shearlet term will become more relevant for the regularization. Evidence of this can be found in [9].

The results in Table 1 show that the nonsmooth formulation outperforms the smooth approach, and this is even more evident as the radius $\mathbf{r}_{\text{ROI}}$ gets smaller. Indeed, when $\sigma = 0.05$, the smooth reconstruction exhibits a 100% ROI relative error for $\mathbf{r}_{\text{ROI}} = 0.1N$, while the nonsmooth reconstruction achieves 24%. This is definitively confirmed by the corresponding images, where no artifacts are visible for the nonsmooth reconstructions, while the smooth ones suffer from low contrast between the smallest features and mild checkerboard effect. Notice that, concerning the figures of merit, the difference between the two approaches is less remarkable when $\sigma = 0.1$, while for the corresponding images (not reported here) similar considerations hold true. However, all the reconstructions are sufficiently good, on a visual basis, since all the fundamental features are detected, and the transition between the ROI to the non-ROI is smooth enough.

For baseline comparison, the analytic approach called Filtered Back-Projection (FBP) with $\sigma = 0.05$ yields PSNR = 27.88, 27.58, 28.49, and rel. err. = 0.54, 1.14, 1.69, respectively. These results are considerably worse than the ones obtained with VMILA. Similar results are obtained with $\sigma = 0.1$. The corresponding images are blurred, suffer from cupping artifacts and the details are not sharp. Some other results, for the smooth case, can be found in [13, 14].

4. Conclusions

In this paper, we presented a numerical assessment on the solution of the ROI CT problem via an iterative minimization method. Two different types of objective functions, namely smooth and nonsmooth, have been considered, aiming at making the reconstruction from truncated data stable. The experiments show that the reconstructions obtained by considering the nonsmooth approach are the best ones, with respect to the figures of merit. A possible explanation might be that the use of the 1-norm leads to the suppression of many small shearlet coefficients in favor of few large shearlet coefficients, that are associated to edges. This allows to separate the structural components of the image from the noise, and this roughly corresponds to denoising. Overall, the reported results show that accurate ROI reconstructions can be obtained regardless of the location and size of the ROI and for rather small ROI sizes using both formulations of the objective function.

Acknowledgments

This work is supported by INdAM-GNCS, the Italian national research project FIRB2012, grant n. RBFR12M3AC, FAR2014-UniFE and FAR2014-UniMORE.

References

[1] Quinto E T 2006 *Proc. of Symposia in Appl. Math.* **63** 1–23
[2] Clackdoyle R and Defrise M 2010 *IEEE Signal Processing* **60** 60–80
[3] Kutyniok G and Labate D 2012 *Shearlets. Multiscale Analysis for Multivariate Data* (Boston, MA (USA): Birkhäuser)
[4] Bonnet S, Peyrin F, Turjman F and Prost R 2002 *IEEE Trans. Image Processing* **11**(3) 169–76
[5] Goossens B, Labate D and Bodmann B 2016 *Private communication*
[6] Bonettini S, Loris I, Porta F and Prato M 2016 *SIAM J. Optim.* **26**(2) 891–921
[7] Natterer F and Wübbeling F 2001 *Mathematical methods in image reconstruction* (Philadelphia, PA (USA): SIAM)
[8] De Man B and Basu S 2004 *Physics in Medicine and Biology* **7** 2463–75
[9] Vandeghinste B, Goossens B, Van Holen R, Vanhove C *et al.* 2013 *IEEE Trans. Nuclear Science* **5** 3305–17
[10] Barzilai J and Borwein J M 1988 *IMA J. Numer. Anal.* **8** 141–8
[11] Frassoldati G, Zanghirati G and Zanni L 2008 *J. Industrial and Management Optim.* **4**(2) 299–312
[12] Lantéri H, Roche M, Cuevas O and Aime C 2001 *Signal Process.* **81** 945–74
[13] Bubba T A, Labate D, Zanghirati G, Bonettini S and Goossens B 2015 *SPIE Optics & Photonics, Wavelets And Applications XVI* vol 9597 (San Diego, CA, USA) p 95970K
[14] Bubba T A, Labate D, Zanghirati G and Bonettini S 2015 *ArXiv e-prints* (*Preprint* 1511.04336)

Journal of Physics: Conference Series **756** (2016) 012010 doi:10.1088/1742-6596/756/1/012010

Phaseless tomographic inverse scattering in Banach spaces

C. Estatico[1], A. Fedeli[2], M. Pastorino[2], A. Randazzo[2], E. Tavanti[2]

[1]Department of Mathematics, University of Genoa, Via Dodecaneso 35, I-16146 Genoa, Italy.

[2]Department of Electrical, Electronic, Telecommunication Engineering, and Naval Architecture, University of Genoa, Via Opera Pia 11A, I-16145, Genoa, Italy.

E-mail: estatico@dima.unige.it

Abstract. In conventional microwave imaging, a hidden dielectric object under test is illuminated by microwave incident waves and the field it scatters is measured in magnitude and phase in order to retrieve the dielectric properties by solving the related non-homogenous Helmholtz equation or its Lippmann-Schwinger integral formulation. Since the measurement of the phase of electromagnetic waves can be still considered expensive in real applications, in this paper only the magnitude of the scattering wave fields is measured in order to allow a reduction of the cost of the measurement apparatus. In this respect, we firstly analyse the properties of the phaseless scattering nonlinear forward modelling operator in its integral form and we provide an analytical expression for computing its Fréchet derivative. Then, we propose an inexact Newton method to solve the associated nonlinear inverse problems, where any linearized step is solved by a L^p Banach space iterative regularization method which acts on the dual space L^{p^*}. Indeed, it is well known that regularization in special Banach spaces, such us L^p with $1 < p < 2$, allows to promote sparsity and to reduce Gibbs phenomena and over-smoothness. Preliminary results concerning numerically computed field data are shown.

1. Introduction

Electromagnetic imaging methods based on inverse scattering analysis [1], especially those working at microwave frequencies, are acquiring an ever growing importance for non-destructive testing in several real applicative fields, ranging from civil, industrial to biomedical engineering [2–6]. Basically, information provided by microwave scattering phenomena measured outside an unknown dielectric object under test are used to characterize the spatial distributions of its dielectric properties (e.g., dielectric permittivity, electric conductivity, and magnetic permeability). The main advantage of microwave imaging is its intrinsic low invasivity with respect, for instance, to X-ray tomographic apparatuses. On the other hand, non-destructive imaging by microwave inverse scattering turns out to be an extremely complicated problem, due to a high non-linear response and a severe ill-positioning of the governing equations that relate the scattered electromagnetic field to the dielectric properties of the scatterer. Hence, regularization methods are required to overcome the ill-posedness and reduce both instability and noise-amplification in the restoration process [7,8].

In conventional inverse scattering, a large amount of information about the scattered field is acquired, that is, a large collection of measured electromagnetic fields by multi-view/illumination and multi-frequency systems, where any measure generally involves both wave magnitude and phase. Although this large set of data can be useful to improve the quality of the whole imaging apparatuses, since it allows for a better restoration of the dielectric properties of the unknown scatterer, its acquisition can be still expensive. A possible way to reduce the whole costs of the imaging apparatuses is to acquire magnitude-only data. Clearly, in this case it is also necessary to employ

[1] E-mail: estatico@dima.unige.it. The work of C. Estatico is partly supported by PRIN 2012 N. 2012MTE38N.
[2] E-mail: alessandro.fedeli@edu.unige.it; matteo.pastorino@unige.it; andrea.randazzo@unige.it.; emanuele.tavanti@edu.unige.it.

inverse scattering procedures that require only information about the magnitude of the scattered field or techniques for approximating the missing information about the phase [9–14].

In this work, a fully phaseless inverse scattering model is introduced and implemented. Basically, neither direct measurements nor indirect estimations of the phase of the electromagnetic scattered field are required to derive the dielectric parameters of the object under test. In this respect, the nonlinear Lippmann-Schwinger integral formulation of the phaseless forward scattering model has been analysed, in order to derive a procedure for its inversion. More specifically, the non-linear phaseless inverse scattering integral equation has been solved by an inexact-Newton method, where the solution of any linearized equation is computed by an iterative regularization algorithm for ill-posed linear functional equations in L^p Banach spaces [15–18]. We recall that regularization in L^p Banach spaces, for $1 < p < 2$, is able to promote sparsity and reduce the over-smoothing effects arising in conventional regularization in Hilbert space L^2. Heuristically, regularization in L^p, with $1 < p < 2$, is based on a convex and smooth minimization which approximates the minimization on the non-smooth L^1 space, which is well-known to be the basic framework of the so-called "sparse recovering". This way, since the L^p-norm is differentiable (as well as the L^2-norm), the computation in L^p, with $1 < p < 2$, is cheaper than dealing with L^1-norm, which is not differentiable, but the sparsity is still promoted [15]. A first numerical validation shows that the proposed low-cost phaseless microwave inverse scattering method in Banach space improves the results of the corresponding phaseless method developed in the standard Hilbert space.

The paper is organized as follows. In Section 2, the considered electromagnetic is described. The developed inversion procedure is detailed in Section 3. Preliminary numerical results are shown in Section 4. Finally, conclusions are drawn in Section 5.

2. The phaseless tomographic microwave inverse scattering model

By assuming cylindrical scatterers and transverse-magnetic illuminations, the electromagnetic scattering phenomena can be described by the following two-dimensional and scalar Lippmann-Schwinger integral equation [1]

$$E_{scat}(\mathbf{r}) = \iint_{D_{inv}} G_0(\mathbf{r}, \boldsymbol{\xi}) c(\boldsymbol{\xi}) E_{tot}(\boldsymbol{\xi}) d\boldsymbol{\xi}, \qquad \mathbf{r} \in D_{obs}. \tag{1}$$

Here D_{obs} is the observation domain where the samples of the electric field are collected, E_{inc} is a known incident electric field, E_{tot} is the total electric field, $c = \epsilon_r - 1$ is the contrast function (being ϵ_r the complex relative dielectric permittivity), G_0 is the free space two-dimensional Green's function [19], D_{inv} is the investigation domain where the unknown scatterer is placed, and $E_{scat} = E_{tot} - E_{inc}$ denotes the scattered electric field. Equation (1) is known as data equation and belongs to the first kind Fredholm integral equations class, hence it is well known to be ill posed. Along with it, we need another one, known as state equation, relating incident and total electric field inside the scatterer

$$E_{inc}(\mathbf{r}) = E_{tot}(\mathbf{r}) - \iint_{D_{inv}} G_0(\mathbf{r}, \boldsymbol{\xi}) c(\boldsymbol{\xi}) E_{tot}(\boldsymbol{\xi}) d\boldsymbol{\xi}, \qquad \mathbf{r} \in D_{inv}. \tag{2}$$

Equations (1) and (2) are discretized by using the method of moments [20]. In particular, D_{inv} is partitioned in N pixels with centers $\mathbf{r}_n^{inv}$, $n = 1, \dots, N$, where both the field and the contrast function are assumed to be constant, and the external electric field is collected in M measurements point $\mathbf{r}_m^{obs}$, $m = 1, \dots, M$, belonging to D_{obs}. Consequently, the following discretized form is obtained (see for example [1] for the details about the involved quantities)

$$\mathbf{E}^{ext}_{scat} = \mathbf{G}^e diag(\mathbf{c})\mathbf{E}^{int}_{tot}$$
$$\mathbf{E}^{int}_{inc} = \left[\mathbf{I} - \mathbf{G}^i diag(\mathbf{c})\right]\mathbf{E}^{int}_{tot} \,. \tag{3}$$

Combining the two equations in (3), we get the following compact matrix equation

$$\mathbf{E}^{ext}_{scat} = \mathbf{G}^e diag(\mathbf{c})\left[\mathbf{I} - \mathbf{G}^i diag(\mathbf{c})\right]^{-1}\mathbf{E}^{int}_{inc} = \mathbf{G}^e diag(\mathbf{c})\mathbf{E}^{int}_{tot}(\mathbf{c}) = \mathbf{F}(\mathbf{c}) \,. \tag{4}$$

This basic formulation involves both amplitude and phase components in $\mathbf{E}^{ext}_{scat}$. However, in this paper we assume that only magnitude measurements are available, i.e., we assume that the data we dispose are only those related to $|E_{scatt}(\mathbf{r})|^2$. Therefore, the scattering model in (4) is modified as follows

$$\mathbf{P}^{ext}_{scat} = \bar{\mathbf{F}}(\mathbf{c}) \circ \mathbf{F}(\mathbf{c}) = \mathbf{F}^{cplx}_{P}(\mathbf{c}) = \mathbf{F}^{cplx}_{P}(\mathbf{Tx}) = \mathbf{F}_P(\mathbf{x}) \,. \tag{5}$$

Here $\mathbf{P}^{ext}_{scat}$ is an array containing the values of $|E_{scatt}(\mathbf{r})|^2$ in the measurement points, the upper bar denotes the element-wise complex conjugate, and $\circ$ denotes the Hadamard product. In order to obtain a real-valued unknown, we rewrite our problem in terms of $\mathbf{x} = [Re(\mathbf{c}^t)\, Im(\mathbf{c}^t)]^t$. Consequently, the transformation $\mathbf{c} = \mathbf{Tx} = [\mathbf{I}\; j\mathbf{I}]\mathbf{x}$ is applied into the model in (5) in order to obtain the final forward operator $\mathbf{F}_p$.

3. The solution of the inverse problem in Banach spaces via quasi-Newton schemes

A successful algorithm to solve the conventional formulation (4) is the inexact-Newton scheme with regularizing Landweber method inside L^p Banach spaces, with $1 < p < 2$ [15]. Due to its flexibility, we choose to apply it also to cope with the proposed phaseless model equation (5). Such approach can be summarized by the following steps (n and l are indexes for outer Newton and inner Landweber loops, respectively):

I) Initialize the procedure with an initial guess $\mathbf{x}_0$. A modified back-propagation procedure is employed to find a suitable non-null value.

II) Linearize (5) around the current estimate $\mathbf{x}_n$. The following linear system is obtained

$$\mathbf{F}'_P(\mathbf{x}_n)\boldsymbol{\delta} = \mathbf{P}^{ext}_{scat} - \mathbf{F}_P(\mathbf{x}_n) = \mathbf{r}_n. \tag{6}$$

Indeed, through the Taylor expansion of (4), that is,

$$\mathbf{F}(\mathbf{c} + \mathrm{d}\mathbf{c}) = \mathbf{G}^e diag(\mathbf{c})\mathbf{E}^{int}_{tot}(\mathbf{c}) + \mathbf{G}^e\left[\mathbf{I} - diag(\mathbf{c})\mathbf{G}^i\right]^{-1} diag(\mathrm{d}\mathbf{c})\mathbf{E}^{int}_{tot}(\mathbf{c}) + O\left[diag(\mathrm{d}\mathbf{c})^2\mathbf{E}^{int}_{inc}\right], \tag{7}$$

the Fréchet derivative of the forward operator in (5) can be explicitly computed as

$$\mathbf{F}'_P(\mathbf{x}) = 2Re\left\{diag\left[\,\bar{\mathbf{F}}(\mathbf{x})\right]\mathbf{F}'(\mathbf{x})\mathbf{T}\right\} \tag{8}$$

where $\mathbf{F}'$ is the Fréchet derivative of the complex-data problem in (4) [16].

III) Approximate the solution of (6) by the following Landweber iterations for the minimization of the residual $\frac{1}{2}\|\mathbf{F}'_P(\mathbf{x}_n)\boldsymbol{\delta} - \mathbf{r}_n\|^2_{L^p}$ in L^p Banach space ($\boldsymbol{\delta}_0 = \widehat{\boldsymbol{\delta}}_0 = \mathbf{0}$)

$$\hat{\boldsymbol{\delta}}_{l+1} = \hat{\boldsymbol{\delta}}_l - \beta \mathbf{F}'^{*}_P(\mathbf{x}_n)\mathbf{J}_p[\mathbf{F}'_P(\mathbf{x}_n)\boldsymbol{\delta}_l - \mathbf{r}_n]$$
$$\boldsymbol{\delta}_{l+1} = \mathbf{J}_q(\hat{\boldsymbol{\delta}}_{l+1}) \tag{9}$$

Here $1 < p < 2$, $\mathbf{F}'^{*}_P(\mathbf{x}_n)$ is the adjoint operator of $\mathbf{F}'_P(\mathbf{x}_n)$, $\beta \in (0{,}2\|\mathbf{F}'_P(\mathbf{x}_n)\|_2^{-2})$ is the relaxation coefficient, $q = p/(p-1)$ is the Holder conjugate of p, $\mathbf{J}_p: L^p \to L^q$ and $\mathbf{J}_q: L^q \to L^p$ are the normalized duality maps of the L^p and L^q Banach spaces respectively. We recall that the normalized duality map $\mathbf{J}_p(s) = \partial\left(\frac{1}{2}\|s\|_{L^p}^2\right)$ of L^p is explicitly computable as $\mathbf{J}_p(s) = \|s\|_{L^p}^{2-p}|s|^{p-1}\mathrm{sign}(s)$, where $\mathrm{sign}(s) = e^{j\arg(s)}$ if $s \neq 0$ and $\mathrm{sign}(s) = 0$ [16] (please notice that, when $p = 2$, then $\mathbf{J}_2(s) = s$).

The iterations are stopped if a maximum number of iterations l_{max} is reached or if the relative variation of the residuals between two consecutive iterations is below a fixed threshold ε_L. Set $\boldsymbol{\delta}$ as the last computed inner iteration $\boldsymbol{\delta}_{l+1}$.

IV) Update the solution with

$$\mathbf{x}_{n+1} = \mathbf{x}_n + \boldsymbol{\delta}. \tag{10}$$

V) Iterate steps II-IV until a maximum number n_{max} of iterations is reached or a predefined stopping rule is satisfied.

The Landweber scheme (9) is proven to be an effective regularization algorithm. We remark that the iterations defined in the equation of (9) are computed in the dual space of L^p [15].

4. A preliminary numerical validation

The developed approach has been preliminarily tested by using numerical data. We focus on a square investigation area of side 1 m, which has been partitioned into $N = 1600$ square cells. The target is a rectangular cylinder with relative dielectric permittivity $\epsilon_r = 2$ and with the dimension shown in Figure 1.

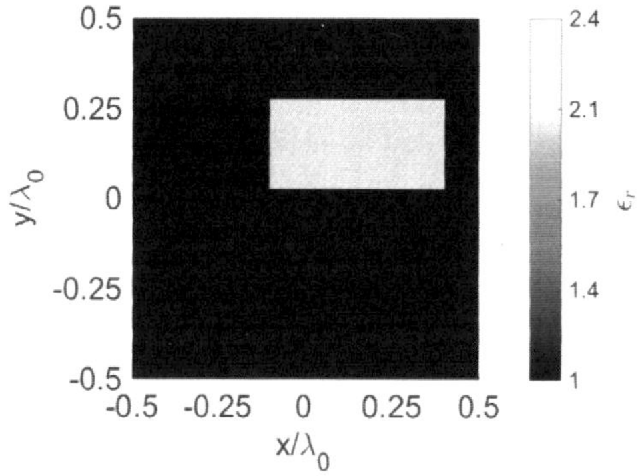

Figure 1. Distribution of the dielectric permittivity in the investigation domain used for tests.

The measurement points are $M = 91$, uniformly located on a circumference of radius 1.5 m and centered in (0,0). The investigation area is illuminated by $V = 8$ line-current sources (the working frequency is 300 MHz). The values of V and M have been chosen to properly sample the field. A zero mean value white Gaussian noise with $SNR = 25$ dB has been added to the simulated field data in order to emulate real operating conditions. The maximum numbers of outer and inner iterations are equal to $n_{max} = 20$ and $l_{max} = 10$, respectively, with $\varepsilon_L = 0.01$. Figure 2 shows the results returned by using $p = 1.4$ and $p = 2.0$; the last one corresponds to the standard Hilbert space inversion.

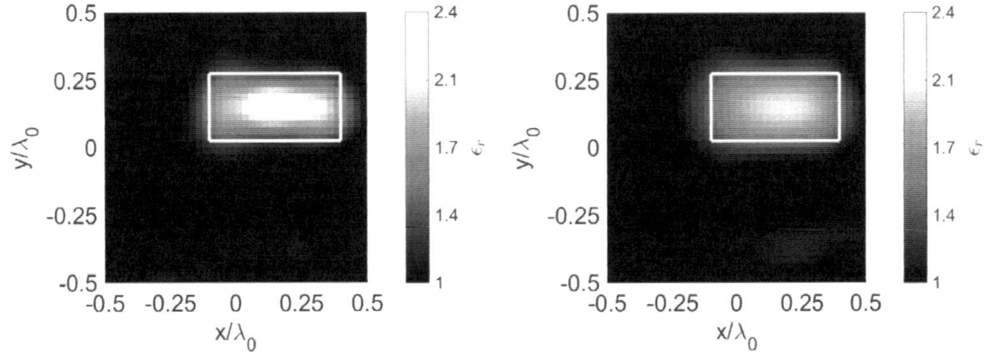

Figure 2. Reconstruction with phaseless measurements. Left: p = 1.4*; Right:* p = 2.0. *The mean reconstruction relative errors on the object are: 0.0731 (left) and 0.1005 (right).*

For $p = 1.4$, the target is located and shaped correctly with a clean background. However, it should be noted that the relative dielectric permittivity of the scatterer is overestimated. On the contrary, for $p = 2.0$, the shape is more irregular and artefacts are present in the background. For comparison purposes, Figure 3 reports the results obtained by using the full complex data (that is, both magnitude and phase) with the procedure developed in [16].

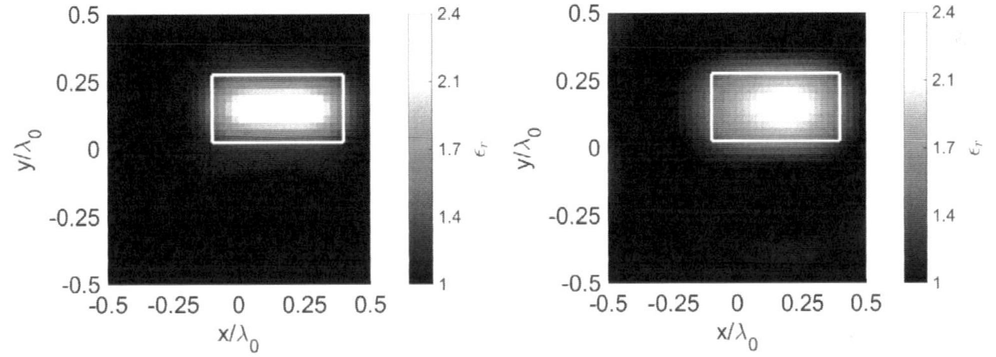

Figure 3. Reconstruction with full complex measurements. Left: p = 1.4*; Right:* p = 2.0. *The mean reconstruction relative errors on the object are: 0.0605 (left) and 0.0692 (right).*

Clearly, the availability of the phase allows a better reconstruction. For $p = 1.4$, the scatterer is again located correctly with a clean background; the overestimation phenomena still persists, although now it is weaker than in the previous case. For $p = 2.0$, considerations about the phaseless case can be proposed again here. We can assert that our phaseless approach exhibits interesting reconstruction capabilities and flexibility due to the tunable p parameters, although the inversion performed with complex (and expensive) measurements has a higher quality.

5. Conclusions

A fully phaseless microwave imaging method has been presented in this paper. The developed approach employs an efficient inversion procedure in the framework of the regularization in L^p Banach spaces, which has been found to provide better results than the corresponding conventional Hilbert space regularization methods. Accordingly, the preliminary numerical validation shows that the proposed phaseless microwave inverse scattering technique in Banach spaces is able to improve the results of the corresponding phaseless method developed in the standard Hilbert space. Clearly, as expected, the method provides worse results than the corresponding approach working with full complex data. However, at least in these preliminary tests, the quality of the restorations obtained by means of the proposed phaseless technique is still quite good. A possible future improvement is to adopt a multi-scaling approach, which is known to be a valid instrument in microwave imaging [21].

References

[1] Pastorino M 2010 *Microwave imaging* (Hoboken, N.J.: Wiley)
[2] Benedetto A and Pajewski L 2015 *Civil Engineering Applications of Ground Penetrating Radar* (Cham: Springer International Publishing)
[3] Hossain M D, Mohan A S and Abedin M J 2013 Beamspace time-reversal microwave imaging for breast cancer detection *IEEE Antennas Wirel. Propag. Lett.* **12** 241–4
[4] Mohammed B J, Abbosh A M, Mustafa S and Ireland D 2014 Microwave System for Head Imaging *IEEE Trans. Instrum. Meas.* **63** 117–23
[5] Pastorino M, Randazzo A, Fedeli A, Salvadè A, Poretti S, Maffongelli M, Monleone R and Lanini M 2015 A microwave tomographic system for wood characterization in the forest products industry *Wood Mater. Sci. Eng.* **10** 75–85
[6] Kharkovsky S and Zoughi R 2007 Microwave and millimeter wave nondestructive testing and evaluation - Overview and recent advances *IEEE Instrum. Meas. Mag.* **10** 26–38
[7] Bertero M and Boccacci P 1998 *Introduction to inverse problems in imaging* (Bristol, UK: IOP Publishing)
[8] Kaltenbacher B 1997 Some Newton-type methods for the regularization of nonlinear ill-posed problems *Inverse Probl.* **13** 729–53
[9] Costanzo S, Di Massa G, Pastorino M and Randazzo A 2015 Hybrid Microwave Approach for Phaseless Imaging of Dielectric Targets *IEEE Geosci. Remote Sens. Lett.* **12** 851–4
[10] Zheng H, Wang M-Z, Zhao Z and Li L 2010 A novel linear EM reconstruction algorithm with phaseless data *Prog. Electromagn. Res. Lett.* **14** 133–46
[11] Lianlin Li, Hu Zheng and Fang Li 2009 Two-Dimensional Contrast Source Inversion Method With Phaseless Data: TM Case *IEEE Trans. Geosci. Remote Sens.* **47** 1719–36
[12] D'Urso M, Belkebir K, Crocco L, Isernia T and Litman A 2008 Phaseless imaging with experimental data: facts and challenges *J. Opt. Soc. Am. A* **25** 271
[13] Caorsi S, Massa A, Pastorino M and Randazzo A 2003 Electromagnetic detection of dielectric scatterers using phaseless synthetic and real data and the memetic algorithm *IEEE Trans. Geosci. Remote Sens.* **41** 2745–53
[14] Franceschini G, Donelli M, Azaro R and Massa A 2006 Inversion of Phaseless Total Field Data Using a Two-Step Strategy Based on the Iterative Multiscaling Approach *IEEE Trans. Geosci. Remote Sens.* **44** 3527–39
[15] Schuster T, Kaltenbacher B, Hofmann B and Kazimierski, K 2012 *Regularization methods in Banach spaces* (Berlin: De Gruyter)
[16] Estatico C, Pastorino M and Randazzo A 2012 A novel microwave imaging approach based on regularization in Lp Banach spaces *IEEE Trans. Antennas Propag.* **60** 3373–81
[17] Estatico C, Fedeli A, Pastorino M and Randazzo A 2015 A multifrequency inexact-Newton method in Lp Banach spaces for buried objects detection *IEEE Trans. Antennas Propag.* **63** 4198–204
[18] Estatico C, Fedeli A, Pastorino M and Randazzo A 2015 Buried object detection by means of a Lp Banach-space inversion procedure *Radio Sci.* **50** 41–51
[19] Pastorino M, Raffetto M and Randazzo A 2014 Two-Dimensional Green's Function for Scattering and Radiation Problems in Elliptically-Layered Media *IEEE Trans. Antennas Propag.* **62** 2071–80
[20] Harrington R 1993 *Field Computation by Moment Methods* (Piscataway, NJ: IEEE Press)
[21] Oliveri G, Lizzi L, Pastorino M and Massa A 2012 A nested multi-scaling inexact-Newton iterative approach for microwave imaging *IEEE Trans. Antennas Propag.* **60** 971–83

Soft Shrinkage Thresholding Algorithm for Nonlinear Microwave Imaging

Hidayet Zaimaga[1], **Marc Lambert**[2]

[1] Laboratoire des signaux et systèmes-CNRS UMR8506, CentraleSupélec-Univ.Paris-Sud, Université Paris-Saclay

[2] GeePs — Group of electrical engineering - Paris, UMR CNRS 8507, CentraleSupélec, Univ. Paris-Sud, Université Paris-Saclay,
Sorbonne Universités, UPMC Univ Paris 06

E-mail: `hidayet.zaimaga@l2s.centralesupelec.fr`

Abstract. In this paper, we analyze a sparse nonlinear inverse scattering problem arising in microwave imaging and numerically solved it for retrieving dielectric contrast from measured fields. In sparsity reconstruction, contrast profiles are a priori assumed to be sparse with respect to a certain base. We proposed an approach which is motivated by a Tikhonov functional incorporating a sparsity promoting l_1-penalty term. The proposed iterative algorithm of soft shrinkage type enforces the sparsity constraint at each nonlinear iteration. The scheme produces sharp and good reconstruction of dielectric profiles in sparse domains by adapting Barzilai and Borwein (BB) step size selection criteria and positivity by maintaining its convergence during the reconstruction.

1. Introduction

Development of efficient reconstruction methods and techniques that exploit sparseness regularized formulations have been widely emerged for solving inverse electromagnetic scattering problems in recent years. High demand of such methods in various applications such as material characterization, subsurface prospecting, remote sensing, and non-destructive testing and evaluation [1, 2] enforces the importance and the need of effective and accurate methods. Inverse electromagnetic scattering problems reconstruct material properties such as permittivity and conductivity in an unknown region from measured electromagnetic fields. However, implementation of such stable, reliable, and efficient reconstruction algorithms is challenging because of the nonlinearity of the scattering equations and ill-posedness of the problem [2–4].

Several approaches have been proposed in order to overcome these issues. Global optimization tools, multi-step information retrieval techniques, and qualitative methods have been introduced to alleviate the non-linearity or its effects. Moreover, first order approximations such as diffraction tomography, Born and Rytov approximations have been proven in order to linearize the problem [2, 3, 5]. On the other hand, innovative sparseness-regularized formulations have recently emerged as an effective recipe to overcome the non-uniqueness and/or numerical instability of the inversion process [6, 7]. The reason behind this is that many images have sparse representations with respect to their expansion basis in the wavelet domain and this yields new developing approaches that minimize the cost functions with zeroth/first norm penalty terms

Journal of Physics: Conference Series **756** (2016) 012011 doi:10.1088/1742-6596/756/1/012011

using highly effective iterative shrinkage thresholding algorithms. Such an increased interest is proven by number of publications in wide domains [4, 6–10].

The aim of the present work is to develop a novel sparse reconstruction algorithm and demonstrate its performance in microwave imaging. In the proposed work sparse reconstruction algorithm of iterative soft shrinkage type is studied. Sparsity constraint is directly applied to the problem of reconstructing the complex internal dielectric properties of an object based on knowledge of the external scattered field which is generated by the interaction between the object and a known incident field. The nonlinear optimization problem is solved by iterative algorithm of soft shrinkage in order to enforce the sparsity constraint by a soft thresholding function. We exploit an adaptive step size selection by BB rule and a priori constraint called as positivity in order to have improved reconstructions. The proposed method provides necessary and sufficient conditions to yield well-posedness and convergence [11–13].

2. Sparse Problem Statement and Discretization

Consider a 2D configuration with transverse magnetic polarization (TM) case where the object under investigation illuminated by a given source numbered as $i, i = 1, ..., Ns$ as in fig. 1. Let E^{inc} be an incident electric field which is generated by a source and polarized along the z-axis with an implicit time factor $\exp(-\mathrm{i}\omega t)$. The object is considered in an investigation domain D and the different media are characterized by their propagation constant $k(\mathbf{r})$ such that $k(\mathbf{r})^2 = \omega^2\varepsilon_0\varepsilon_r(\mathbf{r})\mu_0 + \mathrm{i}\omega\mu_0\sigma(\mathbf{r})$, where ω is the angular frequency, ε_0 and μ_0 are the permittivity and the permeability of the air respectively, $\varepsilon_r(\mathbf{r})$ and $\sigma(\mathbf{r})$ are the relative permittivity and conductivity of the medium as $\mathbf{r} \in D$ is an observation point. The dielectric properties of D are described by the inhomogeneous contrast function, $\chi(\mathbf{r})$, which is defined for non-magnetic area such as $\chi(\mathbf{r}) = \left(k(\mathbf{r})^2 - k_B^2\right)$ where k_B is the propagation constant of the embedding medium.

A known incident field interacts with the scatterer yielding to a total field which is the sum of the incident and scattered fields. Assuming no magnetic media and considering the boundary and radiation conditions we can obtain two coupled contrast source integral equations by applying Green's theorem to Helmholtz wave equations [1, 5] as following

$$E^{\mathrm{tot}}(\mathbf{r}) = E^{\mathrm{inc}}(\mathbf{r}) + \int_D G(\mathbf{r}, \mathbf{r}')J(\mathbf{r}')\mathrm{d}\mathbf{r}' \quad \forall \mathbf{r} \in D \tag{1}$$

$$E^{\mathrm{scatt}}(\mathbf{r}) = \int_D G(\mathbf{r}, \mathbf{r}')J(\mathbf{r}')\mathrm{d}\mathbf{r}' \quad \forall \mathbf{r} \in L \tag{2}$$

where $G(\mathbf{r}, \mathbf{r}') = \frac{\mathrm{i}}{4}\mathrm{H}_0^{(1)}\left(k_B\|\mathbf{r} - \mathbf{r}'\|\right)$, $\mathrm{H}_0^{(1)}$ is the 0-th order Hankel function of the first kind and E^{scatt} is the scattered field. We also defined the contrast source induced within the object by the incident wave such as $J(\mathbf{r}') = \chi(\mathbf{r})E^{\mathrm{tot}}(\mathbf{r}')$, where E^{tot} being the total field in the object. Solving the equations (1) and (2) the unknown contrast function $\chi(\mathbf{r})$ can be determined. Following this, the scattering equations are discretized with the point-matching Method of Moments as in [5] by considering pixel basis functions

$$\psi_k(\mathbf{r}) = \begin{cases} 1 & \mathbf{r} \in D_k, \\ 0 & \text{otherwise}, \end{cases} \tag{3}$$

D_k being the $k-$th pixel and the unknown contrast defined as $\chi(\mathbf{r}) = \sum_{k=1}^{N} \chi_k\psi_k(\mathbf{r})$, N being the number of pixels.

3. Nonlinear Optimization

Soft shrinkage is an approach which minimizes a nonlinear Tikhonov functional with sparsity promoting penalty term. The algorithm is based on the iterated soft shrinkage approach originated for linear operators in the work [6]. A generalization to nonlinear inverse problems has been studied in [4, 14].

The algorithm performs a gradient descent step which involves the adjoint gradient of the cost function with a step size κ, and then a shrinkage step. The latter enforces the sparsity of the reconstruction by setting the small coefficients to zero. The solution of the inverse problem can be obtained by minimizing the cost function which is of the form

$$F(\chi) = \underbrace{\frac{1}{2}\left\|\xi(\chi) - E^{\mathrm{scatt}}\right\|^2}_{K(\chi)} + \alpha\|\chi\|_{l_1} \tag{4}$$

whereas $\xi(\chi) = G(\mathbf{r},\mathbf{r}')\chi(\mathbf{r}')[I - G(\mathbf{r},\mathbf{r}')\chi(\mathbf{r})]^{-1}E^{\mathrm{inc}}$. The l_1 penalty can promote a-priori knowledge of the sparse representation.

Iterative soft shrinkage has the form as (4) where $\xi : X \mapsto Y$ is counted to be a bounded and nonlinear operator with respect to unknown contrast. The algorithm is started by choosing an initial guess χ^1, and continues the iteration as

$$\chi^{j+1} = S_\alpha\left(\chi^j - \kappa\xi'^{*}(\chi^j)\left[\xi(\chi^j) - E^{\mathrm{scatt}}\right]\right) \tag{5}$$

where $\xi'(\chi)$ is the gradient of nonlinear function $\xi(\chi)$, and $\xi'^{*}(\chi)$ is the adjoint operator. S_α is the soft shrinkage operator defined componentwise by [6]

$$(S_\alpha(\chi))_i = \begin{cases} (|\chi_i| - \alpha)\mathrm{sign}(\chi_i), & \text{if } \ |\chi_i| > \alpha \\ 0, & \text{otherwise.} \end{cases} \tag{6}$$

It truncates small values to zero and shrinks large values. The term $\xi'^{*}(\chi^j)\left[\xi(\chi^j) - E^{\mathrm{scatt}}\right]$ is the gradient of the discrepancy $\frac{1}{2}\left\|\xi(\chi) - E^{\mathrm{scatt}}\right\|^2$. The algorithm which has been computed, is as following:

Algorithm 1 Steepest descent reconstruction algorithm with sparsity constraint

Initialize χ^1 and α
for $j = 1, \cdots, J$ **do**
  Solve the direct problem $E^{\mathrm{scatt}}\left(\chi^j\right)$
  Compute the gradient $K'\left(\chi^j\right) = \nabla_\chi K\left(\chi\right)|_{\chi^j}$
  Determine the step size κ_j
  Update inhomogeneity by $\chi^{j+1} = \chi^j - \kappa_j K'\left(\chi^j\right)$
  Threshold χ^{j+1} by $S_\alpha\left(\chi^{j+1}\right)$
  check stopping criterion
end for
output approximate the minimizer of (4).

3.1. Step Size Selection

Selecting the proper step size adaptively is important as it can increase the convergence speed. Therefore, the step size κ can be determined in order to accelerate the algorithm. The iterative

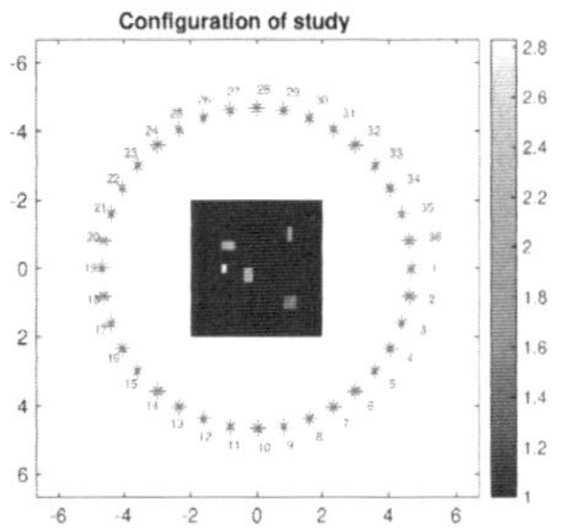

Figure 1: Measured configuration of actual permittivity profile and source-receiver locations on $(x(\mathrm{m}), y(\mathrm{m}))$ axis.

soft shrinkage algorithm with a fixed step size favors the classical Landweber method. Thus, the motivation for increasing the rate of convergence is the comparison with the classical Landweber iteration whose slow convergence results from using a constant step size which is very small [4, 10]. Hence, we select the step size in a way to increase the convergence speed where we only consider the steepest descent operation $\chi^j - \kappa K'(\chi^j)$ of the algorithm. The selection is done by the two-point rule of Barzilai and Borwein which calculates the step size as [15]

$$\kappa_j = \frac{\langle \chi^j - \chi^{j-1}, K'(\chi^j) - K'(\chi^{j-1})\rangle}{\langle K'(\chi^j) - K'(\chi^{j-1}), K'(\chi^j) - K'(\chi^{j-1})\rangle}. \tag{7}$$

3.2. Priori Constraints

Imposing a-priori constraints can sometimes improve the quality of solutions to the inverse problems in a great portion [16]. Towards this end, non-negativity is important in applications like imaging [6, 16]. We know that in order to have a physical solution the unknowns we are dealing with should have constraints, the latter being $\epsilon_r(\mathbf{r}) \geq 1$ and $\sigma(\mathbf{r}) \geq 0$. However, in the general case (as embedded obstacle in half-space) those constraints do not impose non-negativity to the real and imaginary parts of the contrast function. We can consider the constraint by "Projection" where at each iteration the following projection $\epsilon_r(\mathbf{r}) = \max(\epsilon_r(\mathbf{r}), 1)$ and $\sigma(\mathbf{r}) = \max(\sigma(\mathbf{r}), 0)$ is applied.

4. Numerical Results

In the following, the basis functions ψ_k have been chosen as pixel-based. In our example (see fig. 1), we have five objects of physical characteristics described in Tab. 1. Investigation area D is a $l = 6 \times \lambda$-sided square. The discretization size is $n \times n = 80 \times 80$ for the forward problem and $n \times n = 30 \times 30$ for the inverse problem. The number of transmitters and receivers located around the investigation area are 36 evenly distributed on a circle of radius $r = 7\,\mathrm{m}$. The frequency of the transmitters is 200 MHz. The measured field samples are generated by adding 10 dB white Gaussian noise.

Let us defined the relative error norm as a comparison criteria expressed as

$$\varepsilon_r^{\mathrm{err}} = \frac{\|\varepsilon_r^{\mathrm{rec}} - \varepsilon_r\|_2}{\|\varepsilon_r\|_2}, \quad \sigma^{\mathrm{err}} = \frac{\|\sigma^{\mathrm{rec}} - \sigma\|_2}{\|\sigma\|_2}, \tag{8}$$

where $\varepsilon_r^{\mathrm{rec}}$ and σ^{err} are the reconstructed permittivity and conductivity respectively and ε_r and σ the exact ones. The minimum of cost function being the one reached at the end of the process for each α.

One of the key point of such an inversion is the choice of the regularization parameter α in (6). Different tests have been performed in order to evaluate the sensitivity of the choice of

Table 1: Description of the five obstacles, x, y being the coordinate of the center of the obstacle (in m), L_x, L_y its lengths (in m) and ε_r and σ its relative permittivity and conductivity (the latter in $\mathrm{S\,m^{-1}}$)

#	x	y	L_x	L_y	ε_r	σ
1	−1.5	−1.5	0.5	0.5	1.5	0.0022
2	−0.33	0.44	0.5	0.5	2.25	0
3	1.5	−1.5	0.5	0.33	2	0.004 45
4	0	1.5	0.5	0.33	2	0.022 25
5	1	1.33	0.5	0.5	1	0.022 25

regularization on the solution. In our case it can be observed that the higher the value we choose for α the sharper the reconstructions are. However, the choice of α is not arbitrary as can be seen in Fig.2. We can get good reconstruction by choosing α within the range of $\alpha = 1.5 \times 10^{-3}$ and $\alpha = 1 \times 10^{-2}$ for which $\varepsilon_r^{\mathrm{err}}$ and σ^{err} are the smallest as shown fig. 2b and fig. 2c respectively whereas those errors get bigger for larger α.

Moreover, addition of projection to the sparsity reconstruction method improve the quality of the reconstruction even more. Fig. 2 shows us that we can choose the regularization parameter in a wider range when we add projection to our algorithm.

As a complementary validation Figs. 3 and 4 present a comparison of the map of the permittivity and conductivity respectively for various inversion parameters. The influence of the α parameter without any projection constraint onto the permittivity can be shown by comparing fig. 3b and fig. 3c whereas the influence of the projection constraint is presented fig. 3d. Clearer image is obtained by addition of projection constraint with an appropriate α.

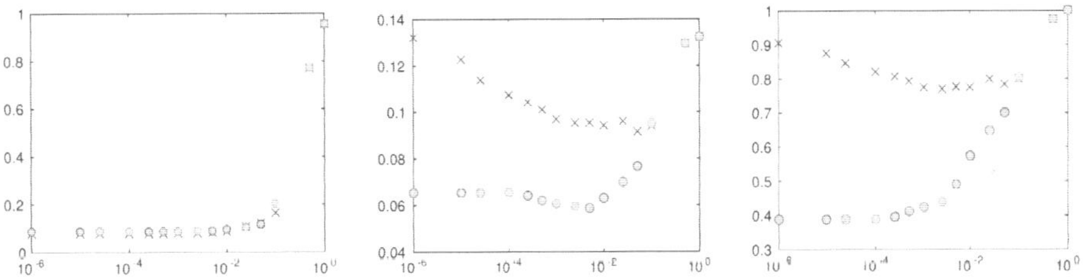

(a) Min. of Cost Function (b) Error for ε_r (c) Error for σ

Figure 2: Error on the cost function (as defined in eq. 4), $\varepsilon_r^{\mathrm{err}}$, and σ^{err} (as defined in eq. 8) as a function of α. –with positivity via projection and × –without positivity

5. Conclusion

Soft iterative thresholding is used to solve the 2-D electromagnetic inverse scattering problem based on l_1 norm penalty term. Retrieval of arbitrary complex shaped targets from simulated data shows that this approach is effective and gives us sharper reconstructions by adopting wavelet basis function. Taking advantage of BB method for step size selection and adding projection enhance the effectiveness of proposed method. Even if choosing the regularization parameter is challenging, addition of projection can lead us to choose it in a wider range possibilities.

Even though the results are promising and advantageous, investigations in experimental settings need to be done to show the feasibility of the algorithm whereas this would be our next step. The development of nonlinear inversion with sparsity regularization for higher permittivity

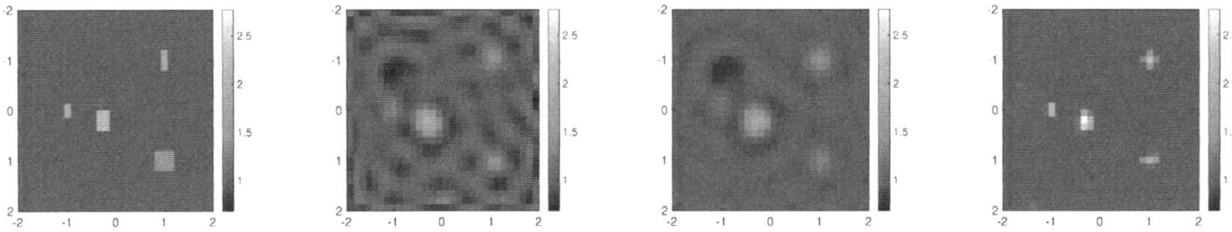

(a) Exact ε_r (b) Without projection, $\alpha = 1 \times 10^{-6}$ (c) Without projection, $\alpha = 5 \times 10^{-3}$ (d) With projection, $\alpha = 1 \times 10^{-6}$

Figure 3: Permittivity ε_r obtained by using sparsity and positivity with 10 dB noise data.

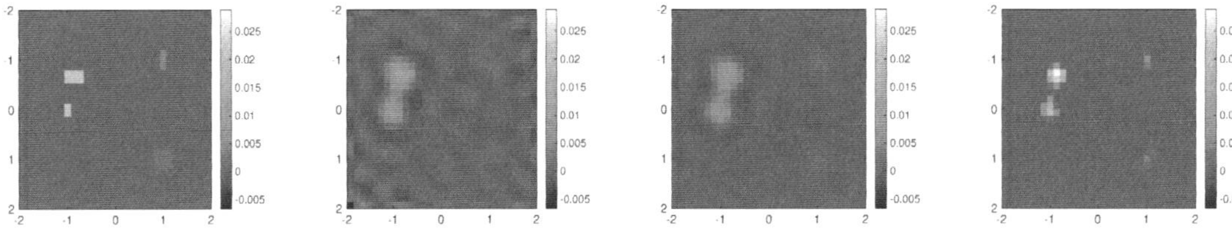

(a) Exact σ (b) Without projection, $\alpha = 1 \times 10^{-6}$ (c) Without projection, $\alpha = 5 \times 10^{-3}$ (d) With projection, $\alpha = 1 \times 10^{-6}$

Figure 4: Conductivity σ obtained by using sparsity and positivity with 10 dB noise data.

values will be studied. Smoothness constraint for the gradient and projection on gradient (as mentioned in section 3.2) will be added to the proposed algorithm in order to analyze the improvement in the reconstruction quality.

Acknowledgment

This work is supported by DIGITEO France through the SIRENA project (2014–2017) under the "Call for Chairs 2014".

References

[1] Pastorino M 2010 *Microwave imaging* vol 208 (John Wiley & Sons)
[2] Poli L, Oliveri G and Massa A 2012 *IEEE Trans. Antennas Propagat.* **60** 2865–2879
[3] Di Benedetto F, Estatico C, Nagy J G and Pastorino M 2009 *Electron. Trans. Numer. Anal.* **33** 105–125
[4] Ramlau R and Teschke G 2006 *Numer. Math.* **104** 177–203
[5] Richmond J H 1965 *IEEE Trans. Antennas Propagat.* **13** 334–341
[6] Daubechies I, Defrise M and De Mol C 2003 *arXiv preprint math/0307152*
[7] Jin B and Maass P 2012 *Inverse Prob.* **28** 123001
[8] Desmal A and Bagci H 2015 *Prog. Electromagn. Res.*
[9] Grasmair M, Haltmeier M and Scherzer O 2008 *Inverse Prob.* **24** 055020
[10] Hanke M, Neubauer A and Scherzer O 1995 *Numer. Math.* **72** 21–37
[11] Bagci H and Desmal A 2014 *IEEE Trans. Antennas Propagat.* **62** 3878–3884
[12] Gehre M, Kluth T, Lipponen A, Jin B, Seppänen A, Kaipio J P and Maass P 2012 *J. Comput. Appl. Math.* **236** 2126–2136
[13] Jin B, Khan T and Maass P 2012 *Int. J. Numer. Methods Eng.* **89** 337–353
[14] Bonesky T, Bredies K, Lorenz D A and Maass P 2007 *Inverse Prob.* **23** 2041
[15] Barzilai J and Borwein J M 1988 *IMA J. Numer. Anal.* **8** 141–148
[16] Vogel C R 2002 *Computational methods for inverse problems* vol 23 (SIAM)